OECD Food and Agricultural Reviev

Innovation, Agricultural Productivity and Sustainability in Australia

OECD
BETTER POLICIES FOR BETTER LIVES

This work is published under the responsibility of the Secretary-General of the OECD. The opinions expressed and arguments employed herein do not necessarily reflect the official views of OECD member countries.

This document and any map included herein are without prejudice to the status of or sovereignty over any territory, to the delimitation of international frontiers and boundaries and to the name of any territory, city or area.

Please cite this publication as:
OECD (2015), *Innovation, Agricultural Productivity and Sustainability in Australia*, OECD Food and Agricultural Reviews, OECD Publishing, Paris.
http://dx.doi.org/10.1787/9789264238367-en

ISBN 978-92-64-23832-9 (print)
ISBN 978-92-64-23836-7 (PDF)

Series: OECD Food and Agricultural Reviews
ISSN 2411-426X (print)
ISSN 2411-4278 (online)

Revised version, January 2016
Details of revisions available at: http://www.oecd.org/about/publishing/Corrigendum-IAPSAustralia2015.pdf

The statistical data for Israel are supplied by and under the responsibility of the relevant Israeli authorities. The use of such data by the OECD is without prejudice to the status of the Golan Heights, East Jerusalem and Israeli settlements in the West Bank under the terms of international law.

Corrigenda to OECD publications may be found on line at: *www.oecd.org/about/publishing/corrigenda.htm*.

Foreword

Innovation, Agricultural Productivity and Sustainability in Australia is part of the series OECD Food and Agricultural Reviews. It was undertaken at the request of the Australian authorities, represented by the Department of Agriculture.

This review examines Australia's policy conditions surrounding business and allowing innovation in the agriculture and food sectors for increased productivity and environmental sustainability. It starts with an overview of the food and agriculture sector and outlines development challenges and opportunities (Chapter 2). A wide range of policies which influence incentives for innovation are then examined: economic stability, governance and trust in institutions (Chapter 3); a favourable and predictable environment for investment (Chapter 4); capacities and public services enabling business development (Chapter 5); agricultural policy (Chapter 6) and the operation of the agricultural innovation system (Chapter 7).

Country policies are analysed following a framework developed by the OECD as part of its work on agricultural innovation and in response to a request from the G20 in 2012 under the Presidency of Mexico, although all aspects may not be fully covered. In this first test phase, the framework has also been applied to Brazil and Canada. Additional countries will be studied in subsequent work and the framework is being continuously revised and improved, in particular to reinforce the coverage of sustainability and structural adjustment issues.

This review was prepared by Olga Melyukhina and Catherine Moreddu. Lihan Wei and Christine Arriola provided statistical support. Hélène Dernis from the OECD Directorate for Science, Technology and Innovation and Douglas Lippoldt, formally of the OECD, provided data and expertise on intellectual property protection. Martina Abderrahmane provided editorial assistance and Michèle Patterson editorial and publication support. This review also benefitted from comments by Ken Ash and Frank Van Tongeren.

The review draws on the responses to the innovation framework questionnaire and the material provided by the Australian Government for the 2011 OECD Conference on Agricultural Knowledge Systems, as well as the work on agricultural productivity by the Australian Bureau of Agricultural and Resource Economics and Sciences (ABARES). In particular, Chapter 2, Chapter 6 and parts of Chapter 7 of this review draw on a recent report by Emily Gray, Max Oss-Emer, and Yu Sheng, Australian Agricultural Productivity Growth: Past Reforms and Future Opportunities (2014). The review also draws on OECD analysis in other economic and social policy fields, and uses cross-country indicators from the OECD and other international institutions.

This review benefited greatly from co-operation with the Australian government. We would like to thank Mathew Worrell, Lee McLean and Wilhelmine Brown from the Australian Department of Agriculture for their continuous support to this project. Thanks are also due to the Department of Environment, the Department of Industry, the Department of Foreign Affairs and Trade, the Department of Employment, the Department of Immigration, the Attorney General's Department, Australian Treasury, and ABARES. We are grateful for their attention to this Review, in particular for most helpful comments on the draft text and additional policy information provided to the OECD Secretariat.

This review was declassified by the Working Party on Agricultural Policies and Market in November 2014.

Acronyms

ABARES	Australian Bureau of Agricultural and Resource Economics and Sciences
ABS	Australian Bureau of Statistics
ACIAR	Australian Centre for International Agricultural Research
APVMA	Australian Pesticides and Veterinary Medicines Authority
ARC	Australian Research Council
AUD	Australian dollar
CGIAR	Consultative Group on International Agricultural Research
COAG	Council of Australian Governments
CPI	Consumer Price Inflation
CPM	Carbon Pricing Mechanism
CRC	Cooperative Research Centre
CSE	Consumer Support Estimate
CSIRO	Commonwealth Scientific and Industrial Research Organisation
EC	Exceptional Circumstance
ERF	Emissions Reduction Fund
FDI	Foreign Direct Investment
FTA	Free Trade Agreement
GCI	Global Competitiveness Index
GDP	Gross Domestic Product
GEM	Global Entrepreneurship Monitor
GSSE	General Services Support Estimate
GST	Goods and Services Tax
ICT	Information and communication technologies
IFC	International Finance Corporation
IGA	Intergovernmental Agreement on National Drought Program Reform
IMF	International Monetary Fund
IP	Intellectual Property
IPR	Intellectual Property Rights
IPRIA	Intellectual Property Research Institute of Australia
MFN	Most Favoured Nation (tariff)

MRRT	Mineral Resource Rent Tax
NCP	National Competition Policy
NDP	National Drought Policy
NDRRA	Natural Disaster Relief and Recovery Arrangements
NPI	National Pollutant Inventory
NRA	National Reform Agenda
NWI	National Water Initiative
OECD	Organization for Economic Co-operation and Development
PDF	Pooled Development Fund
PISA	Programme for International Student Assessment
PSE	Producer Support Estimate
PwC	PricewaterhouseCoopers
R&D	Research and development
RD&E	Research, Development and Extension
RDC	Research and Development Corporation
SME	Small and medium-sized enterprise
SOE	State owned enterprise
STI	Science, Technology and Innovation
TAFE	Technical and Further Education
TFP	Total Factor Productivity
TSE	Total Support Estimate
UN	United Nations
UPOV	International Convention for the Protection of New Varieties of Plants
USD	US dollar
VAT	Value Added Tax
VET	Vocational Education and Training
WDI	World Development Indicators (data base)
WEF	World Economic Forum
WTO	World Trade Organization

Table of contents

Tables

Figures

Executive summary

Australia's agriculture and food industries are well placed to contribute to the economy's future growth given the robust prospects of global food demand and the continuing high international competitiveness of these sectors.

There are, however, important challenges which call for new ways to exploit agricultural resources and human capital. The more than a decade-long decline in agricultural productivity growth needs to be overcome, coupled with the need to accommodate uncertainties about the impacts of climate change and to respond to societal demands in the areas of sustainable development and animal welfare. The agro-food sector, like others, also needs to absorb exchange rate and cost pressures created by the mining boom. To tap additional opportunities of the higher value food segments, Australian agri-businesses need new knowledge and capabilities to seize demand signals and value opportunities, particularly from more affluent consumers in Asian markets.

A fundamental advantage in Australia is its overall policy framework that is supportive to innovation. Australian businesses operate within an open economy with sound macroeconomic fundamentals, flexible labour and capital markets, and competitive product markets. For Australia to retain its position of strong performer requires continuously advancing towards the top levels of global performance in these framework policy areas. Australia's good public governance, including the established process of consultative policy reviews, has traditionally helped it to formulate and to implement desired reforms.

Regulatory burden on farmers could be reduced through greater coherence of regulations across Australian jurisdictions. In particular, further attention could be paid to: commercialisation of genetically modified (GM) crops; registration and approval procedures for national chemicals; requirements for chemicals users; timeliness of import risk analysis for biosecurity; domestic biosecurity and food regulations; and the functioning of water entitlements. Improvements in these areas could facilitate earlier adoption of new technologies and products, easier access to imported inputs to support innovation, and a wider national spread of innovations.

Assisting lenders to evaluate risks in financing innovations could increase access to finance by innovating businesses. Australia has a well-functioning banking sector, with large, sufficiently capitalised and profitable banks serving the agro-food system. However, according to the Australian innovation survey, agricultural businesses identify the lack of additional finance as the most frequent barrier hampering innovation. Although this may reflect appropriate pricing of credit for riskier investment, there may be also a failure of capital markets to cater to innovating borrowers due to a lack of sufficient information and knowledge among lenders to assess the risks involved.

Businesses operate under uncertainty due to transformation of the tax framework. The present government has embarked on a reform of the tax system, including the repeal of the recently legislated Mineral Resource Rent Tax and of other tax provisions; and further broader tax changes are under consideration.

Fostering skills and closing the infrastructure gaps could provide an important boost to innovation. Agriculture is the sector with the highest share of employed who are above the retirement age and its labour is aging more rapidly than elsewhere in the Australian economy. With many farmers nearing retirement and competition from more lucrative activities, the sector may face a tightening labour supply, in particular of skilled farm managers. The uncertainty about future labour supply is compounded by an apparent mismatch between demand for and supply of agro-food skills. The recent infrastructure development notwithstanding, Australia continues to lag behind its competitors in agro-food trade by the availability and quality of its infrastructure. Existing estimates of the agro-food sector's unmet and future infrastructure needs are unclear and require a better understanding of future development patterns. This is important to assist investment decisions in infrastructure by private and public investors.

Agricultural policy is overall conducive to innovation – it creates few market distortions and focuses instead on the sector's long-term development needs. Farming and food industries operate in a market open to international trade and investment. Support is focussed on assisting farmers to manage production risks, encouraging private investments in productivity improvements and sustainable resource use, with limited recourse to measures that distort agricultural prices and costs.

To increase innovation, a refocussing of drought policy from risk coping to preparedness and adaptation should be continued. From the perspective of innovation, risk management policy plays a particular role due to its effects on farmers' investment decisions. The focus on farmers' preparedness and adaptation to climate risks was pledged in the 2013 drought policy reform. It is important to sustain this commitment in the implementation of the new drought policy, which would also go towards creating stronger incentives for farmers to undertake innovation. Beyond on-farm practices, measures to facilitate water trade and to assist in the development of new risk management tools, for example index-based insurance, can enhance farmer's adaptation to climate risks.

The effects of taxation on future structural adjustment and productivity in agriculture are as yet unclear. The current broad tax reforms will likely have a strong impact on agriculture and agro-food industries, due in particular to the easing of taxation for the resource sectors which compete with agriculture for land and labour. The tax reforms may also alter the scope and scale of tax concessions currently available to agriculture and food industries. An assessment of these effects on production and investment in agriculture and the consequent structural adjustment could inform the fine-tuning of policy to improve productivity and sustainability outcomes.

The **Australian rural innovation system is responsive to demands from supply chains**, which drive and co-finance applied research, development and extension (RD&E). This facilitates the adoption of new technologies and management practices that improve farm productivity and the sustainable use of resources at the farm level. As the diversity of the farm population grows, however, diffusion of productivity-enhancing innovation is unequal across farms. It is important to better understand the present motivations and limitations, such as the development of part-time farming, availability of resources and skills, and uncertainties linked to climate change. Investment incentives and technical and managerial assistance have an important role to play in innovation diffusion but it is important for the rural innovation system to address broader productivity and sustainability issues. In particular, climate change will require more fundamental changes than marginal improvements in factor use within current production systems.

Agribusiness currently underinvests in rural R&D despite the estimated large long term returns on investment. This may reflect public overinvestment in applied research with short-term benefits to the industry, and low integration of agri-business in the overall agricultural innovation framework. The public sector continues to dominate rural R&D activities and funding, although co-financing by industry levies generates greater spending capacity for research of practical value for each co-funding supply chain.

Stronger international R&D collaboration and engagement would benefit the sector and society. It would give access to different technologies that could be adapted to Australian conditions and would offer opportunities to pool resources and share knowledge. To help the sector face future challenges, public funding should be sufficient to maintain longer term research capacity that allows the Australian system to respond to future demands, adapt foreign innovations, and cooperate with other countries on common challenges and opportunities.

Policy recommendations encompass the following four key areas:

- **Continue advancing towards the top levels of global performance in the framework conditions for innovation** by further efforts for greater national coherence of regulations that affecting agriculture and agro-industries; continued improvements in the functioning of water markets; exploring possibilities to increase access to finance by innovating farm businesses; and reducing uncertainties about the changes to the tax system.
- **Enhance capacities and services for innovation** by further infrastructure development through better planning, reduced complexity of governance and inter-jurisdictional harmonisation of infrastructure regulations; assisting potential investors to build a better vision of future infrastructure needs; evaluating the long-term impact of public support to the modernisation of irrigation infrastructure on water use in agriculture; enhancing arrangements and programmes to raise skills in agriculture and agro-food industries, and better match supply of skills to industry demand.
- **Maintain the overall agricultural policy conducive to innovation and sustainable farming** by keeping the commitment to focus on farmers' preparedness and adaptation to climate change; investigating the link between increased climate risks and the willingness of farmers to invest; exploring the possible impacts of various drought measures on farmers' risk perceptions and their innovation activity; and assessing the impact of tax reform on productivity structural change, and sustainable resource use.
- **Strengthen direct incentives to innovation in food and agriculture** by providing a long-term vision for investment in the rural innovation system in strategic plans; adapting Research and Development Corporations to improve their responsiveness to cross-commodity issues; considering public funding to innovation, R&D and extension activities not covered by existing supply chains; providing stable support to knowledge infrastructure and long-term projects; making processing industries and retailers an integral part of the innovation system; monitoring investment flows to ensure that funding and activities reflect priorities; reviewing technical assistance and the extension system to ensure adequate public and private supply and access by all farmers; and exploring further opportunities for bilateral and multilateral cooperation in R&D and technology transfer.

Chapter 1

Overall assessment and recommendations

This chapter presents the framework applied in the review to analyse the extent to which Australian policies are supportive of innovation for productivity and sustainability and the findings of the review of a wide range of policies in Australia. In each policy area, it develops specific policy recommendations.

A framework for analysing policies for innovation, productivity and sustainability in the food and agriculture sector

Improvements in agriculture productivity growth are required to meet the growing demand for food, feed, fuel and fibre, and must be achieved sustainably through the more efficient use of natural and human resources. A common finding is that a wide range of economy-wide policies affect the performance of the food and agriculture sectors, and thus need to be considered alongside agriculture-specific policies. Recognising that innovation is essential to improving productivity growth sustainably along the whole agri-food chain, OECD work has focused on the performance of agricultural innovation systems.

The framework used in this report to review Australian policies considers policy incentives and disincentives to innovation, structural change and access to natural resources, which are key drivers of productivity growth and sustainable use of resources (Figure 1.1). The current focus is mainly on agricultural innovation systems. The Oslo Manual defines innovation as the implementation of a new or significantly improved product (good or service), or process, a new marketing method, or a new organisational method in business practices, workplace organisation or external relations (OECD and Eurostat, 2005).

Figure 1.1. Policy drivers of innovation, productivity and sustainability in the food and agriculture sector

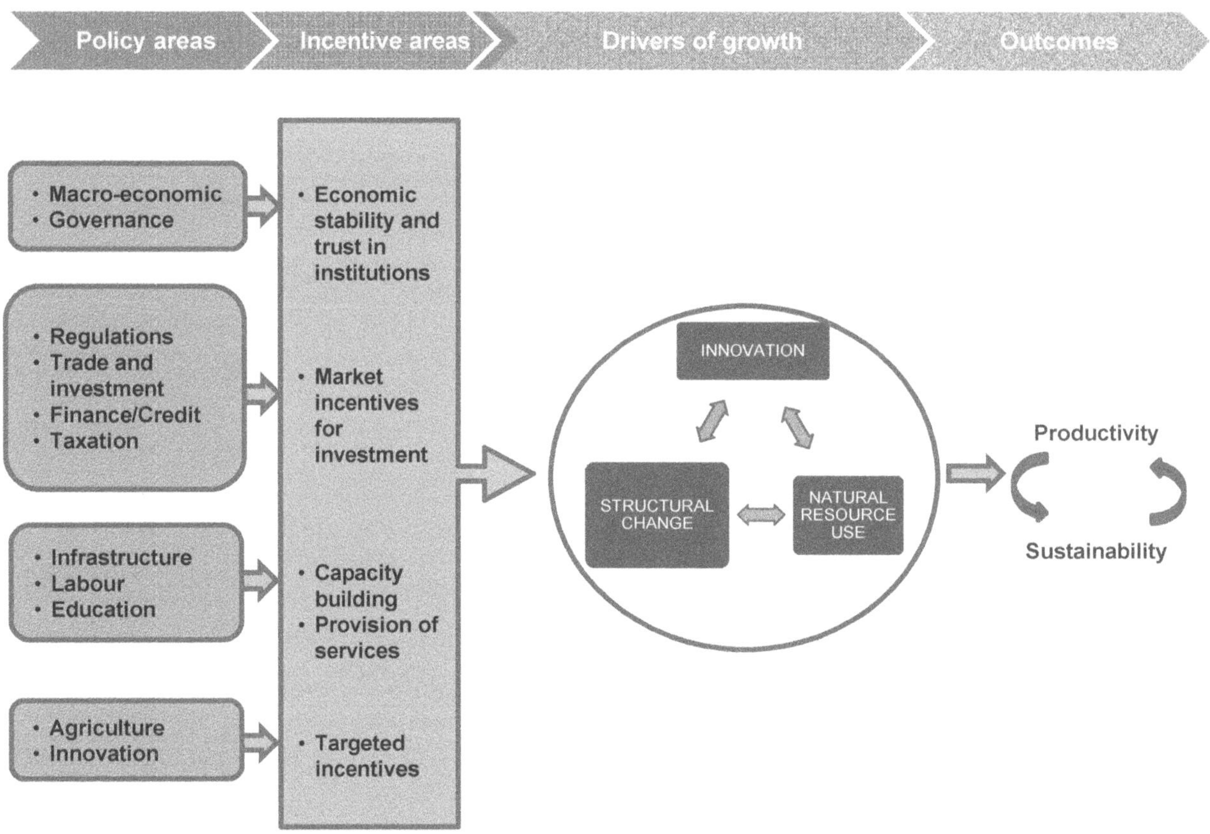

Source: OECD (2014), "Analysing Policies to improve agricultural productivity growth, sustainably: Revised framework", www.oecd.org/agriculture/policies/innovation.

This review begins with an overview of the characteristics and performance of the food and agriculture sector, and outlines the challenges and opportunities (Chapter 2). A wide range of policies is then considered according to the four main channels or incentive areas through which they affect drivers of productivity growth and sustainable use of resources.

- Economic stability and trust in institutions (justice, security, property rights), both of which are essential to attract long-term investment in the economy (Chapter 3).
- Private investment, which in turn requires a transparent and predictable environment that balances the interests of investors and society (Chapter 4).
- Capacity building, including provision of essential public services (Chapter 5).
- Targeted incentives to food and agriculture, which ensure agricultural policies and agriculture innovation systems align the supply of innovation with sector demand and facilitate the adoption of innovation at farm and firm levels (Chapters 6 and 7).

A policy area can affect innovation through more than one channel. Policies can affect innovation positively or negatively depending on the type and intensity of measures. This review reports country-specific information when readily available.

This report aims to review the extent to which the Australian policy environment contributes to improving productivity growth and sustainable use of resources in the food and agriculture sector by fostering the creation and adoption of innovation. Throughout the report, the likely impacts of each policy area on innovation are first discussed in general terms. Specific country measures are then analysed in this regard. Overall assessment and recommendations are drawn from this review on a large range of policy areas.

Challenges to increase agricultural productivity and competitiveness

Australia is in the process of defining its overall development strategy with the view that the country's global competitiveness requires a lift and that new approaches to growth need to be explored. The agriculture and food industries are well placed to exploit growth opportunities given the robust prospects of global food demand and because they are among the most internationally competitive parts of the Australian economy.

The ability to tap additional global food demand largely depends on Australia's success in maintaining productivity growth relative to its trade competitors. Productivity growth in agriculture slowed down considerably in the 2000s, in part due to difficult climatic conditions that prevailed throughout this period. There is also evidence of a declining rate of technical change in agriculture and a widening productivity gap between the front-runner and other farms. The challenge to rekindle agricultural productivity growth is combined with the need to accommodate uncertainties about climate change and to respond to societal concerns on sustainable development and animal welfare, issues which are at present strongly integrated in the domestic political and policy agendas. The agriculture and food industry must also absorb pressures from the mining boom in Australia that increased exchange rate and cost pressures on agriculture, as well as strengthened competition for inputs and factors. To tap additional opportunities in the higher value market segments, Australian agro-businesses require new knowledge and capabilities to seize demand signals and value opportunities, particularly from the promising Asian markets. This also requires a different business model based on a much tighter co-ordination along supply chains. All these challenges call for exploiting agricultural resources and human capital differently, meaning investing in innovative technologies, organisation, and new skills.

Innovation depends on farmers having the incentives and capacity to adopt and implement innovations and being willing to do so. Many inter-related factors determine this process. Innovation is driven by farmer's personal characteristics and depends on specific features of farm businesses, such as the education level, business experience, and income level of the farmer, as well as farm size. There are factors that are beyond the individual farm enterprise that form overall economic incentives and business capacity to innovate. These conditions in turn are influenced by government policies. The focus of this review is on the assessment of policy conditions for agricultural innovation in which businesses invest, innovate and compete. However, ultimately, it is private actors who make decisions to invest in change, responding to a whole set of policy and non-policy factors.

This review comes at a time of extensive policy development initiated by the new Australian government. New policy approaches or broad policy reviews have been announced in key areas, such as competition and deregulation, taxation, labour relations, education, and agricultural policy. Policies and trends discussed in this Review are currently in flux and likely to be overtaken by further developments, thus limiting the scope of recommendations that can be provided at this stage.

Maintain overall supportive policy framework for innovation while moving up to the top levels of global performance

Australia's fundamental advantage is its overall policy framework supportive to innovation

Policy reforms in the 1980s and 1990s contributed significantly to increasing competition and removing constraints to investment and structural adjustment. By a range of international benchmarks, Australian businesses operate within an open economy with sound macroeconomic fundamentals, flexible labour and capital markets, and competitive product markets. The country is one of the world's front-runners for the quality of its public governance and institutions, and its overall regulatory framework is conducive to competition. Australia is one of the most open economies of the world, with few policy barriers to trade and foreign investment, which promotes domestic competition, expands business opportunities and enables transfer of capital, technologies, and expertise to the national economy. Australian businesses are served by a well-functioning and stable banking sector.

For Australia to retain its strong overall framework for innovation involves successful management of the emerging macroeconomic challenges and advancing in areas where there is an apparent scope to catch up with the top levels of global performance. Australia has established mechanisms for broad public policy consultations and comprehensive, independent and transparent policy reviews that help formulate and implement desired reforms.

Economy's rebalancing after the mining boom requires prudent macroeconomic management

A decline in the mining investment boom puts downward pressure on demand, but may be offset by a gradual strengthening of non-mining sectors, including agro-food industries. Macroeconomic policy should allow for a smooth rebalancing of the economy, away from investment in the natural resource sector. The latest *OECD Economic Outlook* emphasises prudence in monetary accommodation that requires timely withdrawal of monetary stimulus. In view of the uncertainties of the post-mining boom adjustment and external demand risks, fiscal consolidation should take better account of structural changes and be more cautious.

Regulatory burden on farmers could be reduced through greater coherence of regulations across Australian jurisdictions

Australian businesses face the least restrictive regulations to start a business among OECD countries, but protection of incumbents and the complexity of regulatory procedures are more restraining than on average across the OECD area. A review of the regulatory burden in the primary sector by the Productivity Commission in 2007 concluded that farmers faced a heavy burden of regulations and identified a set of related business concerns. A follow-up review by ABARES in 2013 noted varying progress in addressing these concerns and areas remain where regulations, if improved, could reduce the burden on rural businesses. This relates to the commercialisation of GM crops; the complexity and timeliness of registration and approval procedures for national chemicals; requirements for chemicals users; timeliness of import risk analysis for biosecurity; the functioning of water entitlements; and domestic biosecurity and food regulations. Improvements in these areas would also facilitate earlier adoption of new technologies and products, easier access to imported inputs to support innovation, and a wider national spread of innovations. The distribution of authority between the Commonwealth and states and territories is distinct and varies across the regulatory issues. Most of the diagnosed issues with the burden on rural business fall under sub-national competency and potential improvements have much to do with achieving better national coherence of regulations across Australian jurisdictions.

Assisting lenders to evaluate risks of financing innovations could improve access to finance by innovating businesses

Australia has a well-functioning banking sector, with large, sufficiently capitalised and profitable banks. For agriculture, borrowing from banks is the principal source of external funding. Following deregulation of the financial sector in the late 1980s, access to bank credit by rural borrowers increased considerably. After a slow-down due to the global financial crisis, agricultural lending seems to be increasing once again. Nevertheless, a survey by the World Economic Forum shows that Australian businesses would like to see simplified procedures for obtaining credit and better opportunities for venture financing. In a 2014 innovation survey by the Australian Bureau of Statistics, agricultural businesses indicated the lack of additional finance as the most frequent barrier hampering innovation. Given the highly competitive nature of rural lending, these results may merely reflect the appropriate pricing of credit for riskier innovative undertakings. However, there may be also a failure of capital markets to cater to innovating borrowers due to a lack of sufficient information and knowledge among lenders to assess the risks involved. This raises the question as to whether there is rationale for government involvement to improve access to finance for innovating farm businesses by addressing, for example, possible information asymmetries amongst rural lenders. Increasing venture finance opportunities for agro-food industries could be an additional venue by identifying constraints to this type of finance and a possible role for the government to ease these constraints.

Businesses operate under uncertainty due to transformation of the tax framework

The previous government implemented large tax initiatives and formulated a broad ranging tax reform agenda. The present government has made the major overhaul of the previous tax system its key economic policy commitment. The recently legislated Mineral Resource Rent Tax has been repealed. This involved the loss of a number of related tax measures as this tax was intended to provide revenue compensating for the effects of other tax reforms. It was also decided not to proceed with a large number of tax measures previously intended for implementation. These and other immediate changes in the tax regime are to be followed by a comprehensive tax reform after delivery of the *White Paper on Tax Reform* in 2015. The importance of tax reforms has been highlighted by the OECD, including the reduction in corporate tax burden, a review of existing tax credits, and the rationalisation of state taxes. It is far beyond the scope of this review to evaluate the recent and potential changes to the tax system. They are likely to have complex trade-offs in terms of business incentives, structural shifts in the economy and environmental sustainability. In the context of this Review of the policy environment for investment and innovation, the observation can be made that frequent and major changes in the tax system create a source of uncertainty for taxpayers and investors. It is important to provide timely signals to business on how the tax reform is likely to develop and to avoid unnecessary delays to legislate the agreed tax changes.

Recommendations on the overall policy framework for investment and innovation

- Continue efforts towards greater national coherence of regulations that affect agro-food businesses in the areas of recognised concern. Ensure these issues are well integrated into the broader work of the Council of Australian Governments that addresses duplication and consistency across Australian jurisdictions.
- Continue improvements in the functioning of water markets, including through harmonisation of water access rights across jurisdictions, enhancing the co-ordination of the system, and its performance assessment.
- Explore the nature of perceived difficulties to access finance by innovating farm businesses. Initiate consultations with lenders to identify areas where information to assess risks involved in financing new agri-business activities and technologies could be improved.
- Raise awareness of farm and agro-food investors about options for non-bank financing. Identify regulatory aspects where incentives to exploit such arrangements could potentially be strengthened and procedures simplified.
- Investigate the demand and supply for venture capital for agro-food industries, particularly for value chain development projects; identify constraints to this type of financing and any possible government role to ease these constraints.
- Provide timely signals to business of intended changes to the tax system and minimise unnecessary delays to legislate the agreed changes.

Infrastructure and skills provision emerge as areas where the potential to boost innovation is large

Reducing infrastructure bottlenecks remains a challenge

Adequate infrastructure is a prerequisite to support the future growth of the agro-food sector by making it more cost-competitive and increasing its capacity to channel products along the domestic supply chain and overseas. Australia has seen significant infrastructure development over the past decade and a half, helped by the broad ranging reforms in this sector and increased investment. Australia, nevertheless, continues to lag behind its competitors in agro-food trade in terms of the availability and quality of infrastructure. Investments in upgrades of irrigation infrastructure have improved water use efficiency, but may also have increased reliance on irrigated agriculture. The existing estimates of agriculture's currently unmet and future infrastructure needs vary, and different sources note that any such estimates require better information and better understanding of future demand for infrastructure. This is important to assist investment decisions in infrastructure by private and public investors. Infrastructure development needs likely differ across key agro-food industries; they also depend on potential changes in the spatial distribution of agriculture due to climate change and the need for the sector to contribute to environmental improvements. According to the Productivity Commission, "institutional and governance arrangements remain deficient and are a major contributor to poor outcomes." Inter-jurisdictional coherence and harmonisation of infrastructure regulations are also the areas where further progress needs to be made.

There are uncertainties about future supply of labour and skills for agriculture

Australia's labour regulation provides a high safety net for employees and allows for a relatively high internal labour mobility. Employment service providers have incentives to increase job placements by attracting those who improve their skills and education. Immigration is an important contributor to Australia's labour force with immigration policy favouring the inflow of skilled labour. The labour regulation framework has recently undergone substantial change with the adoption of the new labour act in 2009. It led to a more employee-friendly system, and its 2012 post-implementation review was generally favourable. However, this act continues to be challenged by business as impeding productivity improvements and has been set for another review by the new government. This signals that the system may be facing changes again with their scope and the degree to which they will affect agro-food industries presently uncertain. It is, nevertheless, worth noting that frequent changes to framework regulations create costs for employers and employees to adjust.

Agriculture is the sector with the highest share of employed who are above the retirement age and with its labour aging more rapidly than elsewhere in the Australian economy. There are suggestions that with many farmers nearing retirement, inter-generational transition may tighten labour supply with a loss of tacit knowledge and experience. There is competition for scarce labour with more lucrative and dynamic sectors, and according to AgriFood Skills Australia, most rural areas struggle to attract young people and retain workers. The uncertainty about the future labour supply in the agro-food industries is compounded with an apparent mismatch between the demand and supply of required skills. Alternative views are that the substitution of capital for labour puts downward pressure on labour demand in agriculture and that the progressive reliance of farm businesses on external providers of knowledge services may compensate for the possible skills gap among farmers. In any situation, the labour system has a role to play in transmitting employer demand for agro-food skills to the education and training systems. It also has a more direct role in contributing to filling the skills gap through job placement and training programmes. Australia's existing general job placement programmes may not be sufficiently focused on the specific needs of the agro-food industry, while current agriculture-specific labour programmes are only narrowly targeted to respond to temporary labour demand peaks. There may be scope for job placement programmes focused specifically on the agro-food industries and which respond to longer-term labour and skill needs of the sector. Similarly, there may be scope for immigration schemes that focus on agro-food industries.

Skills require a push towards top levels of international performance and need to be more strongly guided by industry demand

Australia's education is modern and performing system. The country has an educated population whose attainments and performance are above the OECD average levels, yet it lags behind the best performing countries. OECD skills assessments of both young and adult Australians show that the gap with the best OECD performers is most important in numeracy and mathematics, which are particularly important for technical progress. More than a half of adults show relatively low numeracy proficiency and nearly two-thirds have a limited command of information technology for problem solving. OECD has diagnosed a lagging performance of rural teenagers compared to their peers in non-rural areas. In Australia, the gap with the top five OECD performers is more significant for rural students than it is for non-rural students. Since the late 2000s, the Australian state and territory governments have been active in addressing the skill challenge by setting higher education attainment targets at all levels of education, supported by additional financial incentives and new programmes. Most recently, the present Australian government has announced its reform intentions in education and has launched a number of reviews to inform future decisions. The assessment of the reforms initiated in the late 2000s is difficult due to the short period of their implementation, while the degree to which they will be continued or give place to new approaches is as yet unknown.

Agricultural education is well-funded and offers a variety of courses. At the same time, the number of domestic students enrolled in agriculture and related courses has been decreasing, at least up to 2012. Disciplines directly associated with the agro-food activity, such as soil and agricultural science, animal husbandry, farm management, etc., have the smallest number of enrolments. The decline in student enrolments generally reflects the falling importance of agriculture in the economy. The industry has a principal role in making agricultural and agro-food careers more attractive and it has recently communicated its vision to improve the provision of skills in this sector. This is a broad framework spanning from business capacity to plan jobs and promote skills, through attracting a skilled generation, and to improving and better utilising knowledge and skills of existing workforce. Beyond issues common to many sectors, this framework emphasises the need to tackle the specific challenge of boosting student preference for agro-food careers. It also stresses the importance of exploiting the potential of "on-the-job" training through informal education given the important role of tacit knowledge in agriculture.

Recommendations on capacity for innovation

- Pursue improvements in infrastructure outcomes through more effective planning and reduced complexity of infrastructure governance; continue inter-jurisdictional harmonisation of infrastructure related regulations.
- Undertake a comprehensive assessment of future demands for agro-food infrastructure considering possible climate-related shifts in production pattern, value chain development prospects and environmental targets. Assist potential private and public investors to obtain a better vision of future infrastructure needs.
- Evaluate the long-term impact of public support to the modernisation of irrigation infrastructure on water use in agriculture.
- Explore the feasibility to generate comprehensive data on labour utilisation in the agro-food industries, including temporary and seasonal labour. Undertake a national survey across agriculture and food industries to diagnose current and potential skills bottlenecks.
- Explore arrangements to better match demand and supply of skills in the agriculture and agro-food industries. Encourage industry-government co-operation in the development and updating of training packages, encourage industry-government partnerships in financing education, training, and job placement.
- Explore the scope for job placement programmes and immigration schemes that target agro-food industries and to orient these to longer-term labour and skills needs, thus beyond the currently existing schemes for seasonal labour.
- In the context of fiscal consolidation and educational reforms, maintain the commitment to 2020 education attainment targets.
- Consider a nationally-scoped and co-ordinated campaign for the promotion of agro-food careers that emphasises the opportunities for high-skilled and knowledge-intensive jobs.

Agricultural policy is conducive to innovation, with risk and tax measures to play an important role

Agricultural policy focuses on the sector's long-term development needs

Australia has significantly re-oriented its agricultural policy away from intervening in agricultural markets and subsidising producer costs. This has resulted from the continuous fine-tuning of agricultural policy objectives and instruments to address wider challenges, such as sustainable resource use and productivity improvements, and by targeting assistance to specific outcomes related to these objectives. Total support has been significantly re-allocated to productivity-enhancing general services, such as R&D, infrastructure development, and biosecurity that benefit the farm community and society more generally. Overwhelmingly, policies provide no market price support and domestic prices are largely determined by international markets. Policy support has focused on increasing producer capacity to adjust to and cope with risks, while enabling more sustainable farming. The main instruments include income support payments, concessional loans and grants. These instruments are complemented by a range of tax concessions to agricultural producers.

Keeping the focus of drought policy on farmer preparedness and adaptation will facilitate innovation

In the 2000s, Australian farmers were exposed to significant yield variability which by far exceeded that in European countries; these risks were also systemic, i.e. farmers were affected in many different locations simultaneously. In response to increased income risks, farmers adjusted input use, crop rotations, and product mixes. In some cases, this has enabled a more efficient mix of factors and inputs and productivity improvements, in others, this resulted in sub-optimal combinations of factors and inputs that slowed down productivity growth. The prolonged period of droughts probably also had an impact on farmers' long-term decisions as well and made some reluctant about longer-term and new undertakings, such as investing in new methods and technologies. This could also explain the observed decrease in the rates of technical change and the slower catch-up of other farms which, in turn, led to a slow-down of agricultural productivity growth rates. Australia's agricultural risk management policy plays a particular role in this respect due to its possible effects on farmer willingness to invest.

Providing farmers with the information, knowledge and skills to prepare and adapt to risks helps to expand farmers' planning horizons and thus their willingness and capacity to undertake innovations. This in turn creates a virtuous circle when innovation becomes itself a means to manage production and market risks. In this sense, the pledge of the 2013 national drought programme reform to support self-reliant approaches to business risk management, as opposed to coping with business losses, is conducive to increased innovation.

However, the first drought package for 2014/15 to appear after the adoption of the drought policy reform allocated most of the funding to concessional credit to disaster-affected farmers to cope with the bank debts and cash flow deficits. This has been driven by considerable hardships of the season caused by drought, but there is a longer-run risk of departing from the original reform commitment to focus on facilitating farmers' adaptation. Sustaining this commitment in the implementation of the new drought policy would also go towards creating stronger incentives for farmers to undertake innovation and the sector to adjust. Beyond on-farm practices, measures to facilitate water trade and to assist in the development of new risk management tools, for example index-based insurance, can enhance farmer's adaptation to climate risks.

The effects of taxation in agriculture require better understanding

The use of various tax concessions is a prominent feature of Australia's agricultural policy. Tax incentives are available to farmers in the same way as to all businesses, or provided to them as primary producers or as SMEs. Some tax concessions are aimed at smoothing farm income fluctuations, others to facilitate exit from farming or entry into new farming activities, or adopt good environmental practices. The overall tax reform that is currently under development may change the scope and scale of these tax concessions. Beyond that, agriculture may experience broader impacts from the tax reform, in particular from the removal of the Minerals Resource Rent Tax and the change in greenhouse gases emission policy which are likely to boost production incentives in the resource sectors which compete with agriculture for land and labour. An assessment of these tax effects on production and investment decisions in agriculture and the consequent structural adjustment could inform the fine-tuning of policy to productivity and sustainability outcomes.

Recommendations on agricultural policy

- Keep the commitment to focus on measures to improve the preparedness of farmers and their adaptation to climate change in the implementation of the Intergovernmental Agreement on national drought program reform.
- Investigate the link between the increased climate risks and the willingness of Australian farmers to invest; explore the possible impacts of various drought measures on farmers' risk perceptions and innovation activity.
- Assess the impact of tax reforms, including changes in tax concessions available to agriculture, on productivity structural change, and sustainable resource use.

The rural innovation system is responsive to short-term primary industry demand, but also needs to address broader long-term challenges

The Australian rural innovation system has been successful in responding to the demand for new technologies and management practices of primary agriculture. The close integration between research, development and extension activities has also facilitated the diffusion of innovations at the farm level, which has contributed, together with structural adjustment, to improvements in total factor productivity and sustainability.

A unique feature of the Australian rural innovation system is the Rural Research and Development Corporation (RDC) model of co-financing of rural R&D activities established in 1989. It places interactions between public R&D and agricultural industries at the heart of the rural innovation system and has channelled in recent years a significant share of Australian Government spending on rural research and development (R&D). By matching public funds with producer levies, this model generates greater spending capacity and ensures that producers who benefit from research contribute to its costs; it also ensures that research is of practical value. R&D activities within RDCs involve various rural and general R&D institutions and are organised by commodity, although recently some RDCs address broader challenges at supply-chain level. Originally competitive- and market-driven, the model has become more collaborative and inclusive. However, it does not directly integrate agribusiness processing and retailing stakeholders in funding decisions, potentially limiting capacity to respond to demand for product and process development along the food chain. Similarly, its design is more adapted to marginal improvements than fundamental changes in production systems and resource management.

Governance mechanisms ensure the coherence of a complex and interactive system, and facilitate continuous policy adaptation

The rural innovation system includes a wide range of public and private actors, and funding mechanisms. General innovation institutions, such as the Commonwealth Scientific and Industrial Research Organisation (CSIRO) and universities are the main providers of rural R&D, including through participation in the RDCs. The Cooperative Research Centres (CRCs) Programme, which features partnerships between research funders, suppliers and end-users, often with international partners, and targeted to specific issues, are used to advancing R&D on topics related to agriculture, fisheries and forestry. Agricultural CRCs demonstrated a large return on investment in a recent programme review.

Strong governance and clear rules aim to ensure the coherence of this relatively complex system of actors and institutions, which are jointly developing a National Primary Industries 'RD&E' (Research, Development and Extension) Framework to support greater collaboration and co-operation. It remains nevertheless difficult at present to trace the flow of resources and outcomes, and thus to evaluate the overall system.

The rural system is integrated in the general R&D system. Rural strategic priorities are aligned with national priorities, although there may be delays in adjustment. Priorities for investment in rural R&D are communicated clearly in strategic documents. Evaluations are conducted by government bodies, such as the Productivity Commission, but also by independent reviewers. They include estimates of public and private returns on investment, and opinion surveys of stakeholders. Regular evaluation of programmes has contributed to refinements over the years and efforts are on-going to improve information and analytical tools to evaluate the innovation system nationally, as well as to supply the information needed for strategic investment decisions and performance evaluation.

Public funding of rural R&D has decreased, while demand for innovation has broadened and longer-term challenges receive increasing public attention

The public sector continues to fund rural R&D activities to a large extent (three-quarters). Despite the gradual reduction in public funding for rural R&D, public research intensity (expenditure as a share of agricultural GDP) is relatively high by international standards. In a recent public inquiry, stakeholders, however, pointed to the challenge of meeting broader objectives, such as resource management or innovation along the food chain, with lower public funding. More generally, they questioned the ability of the system to sustain long-term research capacity to respond to future demand, to adapt foreign innovations, and to co-operate with other countries.

Over the years, the fine-tuning of R&D funding mechanisms has aimed to increase the net pay-off of public funds to the community and to ensure they generate additional R&D and associated spill-overs that are worth more to society than a programme's full costs.

The RDC model, in particular, has been continuously fine-tuned to improve the balance between public and private funding, and market-driven and public good activities. In its 2011 review of RDCs, the Productivity Commission argued that the government's funding contribution is likely to have induced only a modest overall amount of additional, socially valuable research. The Productivity

Commission's most recent evaluation of the RDCs suggests that current arrangements involve highly complex and unclear funding flows, that there is inadequate data to perform a comprehensive evaluation of the system, and that despite the existence of various sets of priorities to guide R&D investment, there is inadequate research on aspects that are most beneficial to society as a whole (as opposed to industry interests). It is not clear whether public focus on R&D that generates public benefits, rather than short-term private benefits, will affect public funding to RDCs.

Agribusiness underinvests in rural R&D despite estimated large returns on investment

Through the RDC co-financing model, industry levies dominate private funding but there seems to be underinvestment by agri-business. According to the 2011 Productivity Commission review, this may be due to the cost of doing research locally and the small size of the market, which makes it difficult to develop new markets and products, but also to regulatory burdens and inconsistencies, and particular features of intellectual property protection as applied to rural R&D. The fact that agribusiness processing and retailing are not an integral part of the overall framework is also a problem. For example, submissions to the Productivity Commission inquiry into RDCs suggested there was little consultation with private companies or individual producers as part of the development of the national framework.

Some critics have also argued that high levels of government investment in applied research with short-term, private benefits are actually crowding-out private sector investment. Much of this argument is based on the system's high rates of return. Returns of 15% to 20% should attract more investors than happens in practice.

Stronger international R&D cooperation would benefit the sector and society

The Australian agricultural sector is highly dependent on foreign innovation: the contribution of foreign R&D to Australian agricultural productivity growth is double that of domestic R&D. Given Australia's small domestic capacity for R&D and its small home market, realising benefits from international collaborations and research spill-overs remains a priority. With the exception of its grain sector, Australia does not seem to have a strong history or capacity to cooperate widely with international organisations or researchers in other countries. International collaboration is mainly through ACIAR programmes, collaboration with CGIAR centres, and more recently participation in international networks, including the Global Research Alliance on Agricultural Greenhouse Gases launched in 2009. Stronger international collaboration and engagement would offer significant benefits. It would give access to different technologies that could be adapted to Australian conditions and would offer opportunities to share knowledge to respond to global challenges.

Further productivity gains require wider diffusion of innovation across farms

While innovation has driven past farm performance, productivity growth has been slowing down over the last 15 years despite the applied and responsive nature of the R&D system. As diversity within the farm population is growing, diffusion of productivity-enhancing innovation is unequal across farms. It is important to better understand motivations and limitations, such as part-time farming, lack of resources and skills, the high cost of technology in a small market, and the uncertainties linked to climate change.

Technical and managerial assistance have an important role to play in the diffusion of innovation. The supply of extension services in Australia is diverse, with little direct government involvement except for biosecurity and public good areas such as natural resource management and sustainable production practices. Public extension activities also focus on gathering, interpreting and communicating information on the benefits of the technologies that are available. Public service delivery is through group services rather than individual farm visits. With public services withdrawing from extension, farmers now pay farm management consulting firms for advice. Input suppliers continue to play an important role, which reflects their interest. Public funding of extension services is through programmes, in particular RDCs. Such services often reflect the production-orientated nature of Australian RDCs.

And will depend on the capacity of the system to respond to a changing and more diverse sector

Future agricultural productivity growth will depend on the capacity of the rural innovation system to supply innovations to a diverse sector at a reasonable cost. It is important for the rural innovation system to develop a range of solutions adapted to the diversity of farm characteristics beyond the demands expressed by the supply chains, and addressing longer-term and broader productivity and sustainability issues. There is also the challenge to allow space for innovation that takes a whole farm or land management approach; for innovation in new products adapted to changing natural conditions and growing market demand for higher quality and value-added products; and for more fundamental changes in production systems to adapt to climate change and water scarcity.

Given expanding and competing demands for scarce public funds, a key challenge lies in maximising the payoffs to public investments and to the wider community, while minimising transaction costs across the multiple R&D and extension providers and jurisdictions that comprise the Australian system. Increased co-operation with other countries will also be a key challenge.

Recommendations to strengthen direct incentives to innovation

- When revising government priorities and reviewing funding mechanisms, provide a long-term vision for investment in the rural innovation system.
- Maintain the focus on climate change as a key factor of future competitiveness of Australian agriculture. Continue research and evaluation of the impacts of climate change on agriculture; encourage long-term vision in the innovation agenda based on a comprehensive national adaptation strategy.
- Implement and, if needed, adapt the National Primary Industries RD&E Framework to support greater collaboration and cooperation.
- Adapt RDCs to improve their responsiveness to cross-commodity issues, for example by creating a cross-sector thematic RDC, or by broadening the mandate and partnership of existing ones.
- Consider public funding to innovation, R&D and extension activities not covered by existing supply chains, to respond to demand from a growing diversity of stakeholders in the food and agriculture sector, with different capacity for innovation, and to support the generation of solutions to allow the sector to become more resilient to longer-term sustainability issues such as water variability.
- Provide stable support to knowledge infrastructure and long-term projects, to strengthen the capacity for collaboration at the international level and to allow for break-through innovations.
- Enhance the involvement of processing industries and retailers in innovation, by making them an integral part of the system, from the priority setting stage to the financing and commercialisation of innovation stage.
- Generate data on the adoption of innovation in farm surveys in order to analyse the characteristics of innovators and the main barriers to adoption of innovation. Monitor investment flows to ensure that funding and activities reflect priorities.
- Review technical assistance and the extension system to ensure adequate public and private supply and access by all farmers.
- Explore further opportunities for bilateral and multilateral cooperation in R&D and technology transfer.

References

OECD (2014a), "Analysing Policies to improve agricultural productivity growth, sustainably: Revised framework", available at: www.oecd.org/agriculture/policies/innovation.

OECD and Eurostat (2005), *Oslo Manual: Guidelines for Collecting and Interpreting Innovation Data, The Measurement of Scientific and Technological Activities*, 3rd Edition, OECD Publishing, Paris.
DOI: http://dx.doi.org/10.1787/9789264013100-en.

Chapter 2

Overview of food and agriculture challenges and performance in Australia

This chapter outlines the main challenges and opportunities for the food and agriculture sector of Australia, which will require innovation. It describes the overall economic, social and environmental context in which the sector operates, and the natural resource base upon which it relies. It provides an overview of the general geographical and economic characteristics of the country; and outlines the contribution of the agri-food system to the economy. It identifies the main structural characteristics of primary agricultural and upstream and downstream industries; describes the main food and agriculture outputs and markets; and reviews trends in agricultural productivity and sustainability.

The statistical data for Israel are supplied by and under the responsibility of the relevant Israeli authorities. The use of such data by the OECD is without prejudice to the status of the Golan Heights, East Jerusalem and Israeli settlements in the West Bank under the terms of international law.

Challenges and opportunities: The need for innovation

Australia is developing its new long-term development strategy. The slow-down of productivity growth of the overall economy, a less dynamic demographic outlook, and the societal demand for greener growth create an understanding that Australia's global competitiveness requires a new lift and new approaches to growth need to be exploited. Agriculture and food industries are well placed to respond to growth opportunities, both given the robust prospects for global food demand and because these sectors are among the most internationally competitive parts of the Australian economy.

Future growth in global food demand presents a major opportunity for Australia as a competitive agricultural exporter. This is further supported by the fact that a large part of that growth is projected to come from the Asian region, Australia's close and established market. The country's ability to capture this additional demand will depend on the success of maintaining its competitiveness and productivity growth relative to its trade competitors. Australia showed high productivity growth in agriculture during the 1990s and out-performed large agricultural exporters such as the United States, Canada, and Brazil. In the 2000s, productivity gains slowed down significantly. This is in part attributed to poor climatic conditions during most of this period, but other factors may have also been at play. Recent analysis points to a declining rate of technical change in some industries over the last decade and a half, as well as an increasing gap between the "best performing" and "average" farm, indicating that some Australian farmers may not undertake sufficient innovation.

A challenge to rekindle agricultural productivity growth is combined with the need to accommodate uncertainties about climate change and to respond to societal expectations on sustainable development and animal welfare. These issues are of prominent public concern and are strongly integrated into the political and policy agendas. Agriculture's environmental footprint remains significant and climate change is estimated to produce one of the largest negative impacts on agricultural productivity worldwide. Addressing broader societal concerns means that traditional productivity-enhancing technologies and practices need to be constrained or give place to different ways of exploiting agricultural resources. To reconcile productivity growth and the need to conform to broader societal values, investment in innovative technologies and new skills are required.

The mining boom in Australia has increased competitive pressure on the agricultural sector. A strengthened exchange rate and historically high national terms of trade increase the cost of agricultural exports. Domestic cost pressures are created by competition from the mining and other trade-oriented economy sectors for inputs and factors, such as land and labour. Although mining investment has slowed down recently, resource sectors are expected to remain the most rapidly expanding part of the economy. Agriculture will thus continue to face competition for resources and its ability to attract them will depend on the extent to which it will be able to generate competitive returns.

The ageing population and urbanisation affect Australian agriculture possibly stronger than any other industry relying for labour on rural and regional population. Agriculture is the sector with the oldest age structure. With many farmers nearing retirement, inter-generational transition and competition from more lucrative activities may involve a tightening of supply of labour and skills in the sector. Innovation will facilitate a shift towards less labour-intensive production systems, while technologically advanced systems may become more attractive to younger labour and to talent.

To tap additional opportunities in the higher value market segments, Australian agro-businesses require new knowledge and capabilities to seize demand signals and value opportunities, particularly from the promising Asian markets. This requires a different business model based on much tighter co-ordination along the supply chains.

General context: Natural endowment and economic development

Australia is the sixth largest country by land area, but with an almost 24-million population it is one of the least and unevenly populated countries in the world (Table 2.1). Agricultural land occupies over half of the country's territory and per capita availability of arable land is the highest in the world. Yet despite abundant land, agricultural productive potential is constrained by poor soils and the most variable climate conditions in the world. Desert or semi-arid areas occupy by far the largest part of the territory, while coastal areas are prone to floods. Total freshwater resources are relatively limited and rainfall is extremely variable, leading to major flooding alternating with extended periods of drought. Climate change is of considerable concern: current predictions point to a warming of Australia's climate with more heat and fewer cool extremes. In southern Australia, average rainfall is projected to decrease, while most parts of the country will see heavier rainfalls (Australian Bureau of Meteorology, 2014).

Australia is one of the most advanced world economies. In 2012, it ranked 12th by the size of its GDP and 11th by GDP per capita (in PPP terms); it ranked second in the UN's Human Development Index. It is the country with the second largest estimated wealth per adult (USD 402 600) and the highest median wealth (USD 219 500) (Credit Suisse, 2013). Annual disposable household income is considerably above the OECD average. However, the relative poverty rate (percentage of persons living with less than 50% of median income) and income inequality also exceed the OECD average (OECD, 2014).

Table 2.1. Contextual indicators for Australia

	GDP	GDP per capita	Population	Total land area	Agricultural land	Arable land per capita	Freshwater resources	Freshwater resources per capita m^3
	Billion USD (2012)	PPP USD (2012)	Million (2012)	'000 km^2 (2012)	'000 Km^2 (2011)	Hectares (2011)	Billion m^3 (2011)	m^3 (2011)
Australia	**1 532**	**44 407**	**23**	**7 682**	**4 097**	**2.14**	**492**	**22 039**
Brazil	2 253	11 716	197	8 459	2 750	0.37	5 418	27 512
Canada	1 821	41 150	35	9 094	626	1.25	2 850	82 647
China	8 227	9 059	1 377	9 327	5 191	0.08	2 813	2 093
European Union	16 687	34 064	504	4 238	1 879	0.21	1 505	2 963
India	1 842	3 813	1 237	2 973	1 798	0.13	1 446	1 184
Indonesia	878	4 876	247	1 812	545	0.10	2 019	8 281
Russia	2 015	23 501	143	16 377	2 153	0.85	4 313	30 169
South Africa	384	11 255	52	1 213	964	0.24	45	886
United States	16 245	57 869	314	9 147	4 113	0.51	2 818	9 044

Source: OECD stat; WDI (2013), *World Development Indicators Database*. World Bank, http://data.worldbank.org/data-catalog/world-development-indicators.

Australia has enjoyed uninterrupted economic growth for over two decades, one of the longest periods of sustained growth among OECD economies. The economy expanded at above the OECD average rate (Figure 2.1). Economic growth since the mid-2000s was largely driven by a rising demand for commodities from Asia, in particular China. The global commodity price boom led to increased incomes and investment in the mining and associated sectors, it strengthened the domestic currency and changed the composition of demand. These trends amplified growth disparities in the non-mining sectors, in particular those that were trade-oriented such as manufacturing, agriculture, tourism and education, and placed these sectors under adjustment pressures. The mining investment boom is declining, as evidenced by a recent slow-down in mining investment, fewer jobs created in this sector and, the decrease in the terms of trade. Nevertheless, Australia is well placed to continue benefitting from the growth in Asian countries, particularly China, given the complementarities between the two economies. The degree to which these benefits will be realised largely depends on the future growth in the Asian region and commodity price developments (OECD, 2012a).

Figure 2.1. Australia's real GDP growth in international comparison

At purchasing power parity, 2000=100

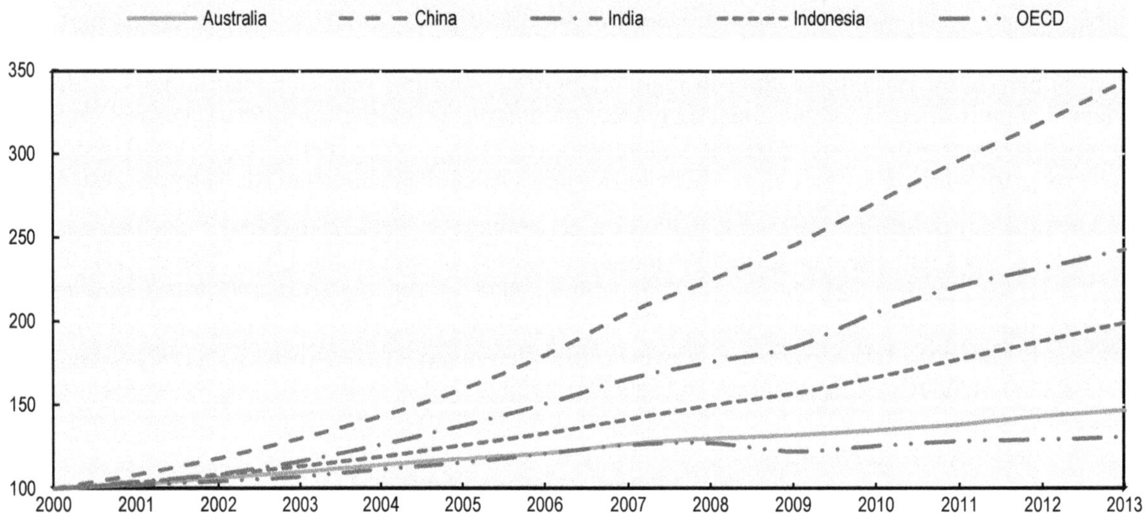

Source: OECD Economic Outlook 96 database - November 2014 http://dotstat.oecd.org.

StatLink http://dx.doi.org/10.1787/888933249671

Productivity gains are key to future broad-based growth. Australia saw a sharp downturn in overall productivity growth over the past decade. The Australian Bureau of Statistics (ABS) estimates that after a peak in the 1990s, multi-factor productivity (MFP) growth gradually fell to zero in the last complete business cycle (2004 to 2008) and has continued to decline since then. The majority of the 12 industries constituting the market sector experienced declines in MFP growth rates during this last business cycle, with the most significant deceleration observed in the manufacturing, mining, and agriculture sectors. There have been sector-specific factors that have contributed to the aggregate productivity slow-down. The mining sector, estimated to have significant effect on the overall dynamics of productivity, saw massive injections of capital and labour, and lags between investment and full capacity may imply a sharp decline in the measured productivity. In the agriculture sector, productivity performance was affected by recurrent droughts (discussed below). Beyond the mining boom and factors that may be temporary or confined to specific sectors, analyses point to the possibility of more systemic reasons for productivity slow-down. One is that the spike in MFP in the 1990s that followed the structural reforms in the 1980s and 1990s may represent a level shift with a catch-up phase, rather than an increase in the long-term growth rate. Another possible factor is that broad-based economic prosperity and soaring profits have likely eased pressure on firms to improve efficiency. Other arguments are that reform has lost momentum, while some new regulations have been "productivity stifling", and that following the long expansion of its economy, Australia now faces capacity constraints, such as skill shortages and infrastructure bottlenecks. The current slow-down may be partly reversed when key resource projects and other infrastructure investments come on stream. Overall, there is much uncertainty about the interpretation and the underlying causes of current productivity trends in Australia, as well as to how these trends are going to project into the future (OECD, 2012a).

The importance of agriculture in the economy[1]

Agriculture is a small but important part of Australia's economy. It accounts for 2.4% of GDP and absorbs the same share of total employment (Table 2.2). Considering non-metropolitan areas, agriculture's role as an employer is important, in particular for remote areas where it provides between 10% and 15% of employment. The sector also contributes to employment in related industries; the food product manufacturing industry employed 200 000 people in 2010/11, more than any other manufacturing industry in Australia (ABS, 2012). The sector's export share is around five times its contribution to GDP, with agro-food exports making up over 10% of all exports of goods and services in 2011/12 (ABARES, 2012). Agriculture's indirect effects on the economy are also important. For example, a widespread drought in 2006–07 is estimated to have decreased economic growth across Australia by around 0.75 percentage points (Penm and Glyde, 2007).

Agriculture is highly important from the perspective of natural resource use and the environment and is a major user of land. Compared to countries that use similar production technologies, Australian agriculture relies heavily on land as an input, using around ten times as much land per unit of output compared with, for example, the United States (Nossal and Sheng, 2013). Agriculture is also the largest consumer of freshwater, absorbing 52% of total water withdrawals in 2008-10 (OECD, 2013). The substantial use of irrigation in parts of the country places pressure on ecosystems that rely on the nation's water (DFAT, 2012). Agriculture is a major source of greenhouse gas emissions, accounting in 2010 for over one-half of methane and over 80% of nitrous oxide emissions (WDI, 2014).

Table 2.2. Importance of agriculture in the economy, 2011

Percentage

	Gross Value Added	Employment	Exports	Imports	Total land area	Total water withdrawals[1]
Australia	**2.4**	**2.4**	**13.1**	**4.8**	**53.3**	**52.2**
Brazil	5.2	17.0	31.9	4.8	32.5	54.6
Canada	1.9	2.0	9.2	7.0	6.9	..
China	10.1	34.8	2.3	5.1	55.7	64.6
European Union	1.7	4.6	6.5	5.9	44.3	..
India	17.4	47.2	..	..	60.5	90.4
Indonesia	14.7	35.8	21.0	10.7	30.1	81.9
Russia	4.3	7.9	1.7	12.2	13.1	19.9
South Africa	2.4	5.1	7.7	6.3	79.4	62.7
United States	1.2	1.6	9.8	4.7	45.0	40.2

.. not available. Australian data on the share of agriculture in gross value added and employment corresponds to 2012 and 2013 respectively, and on the share in freshwater withdrawals corresponds to the average for 2008-10.

1. Latest available year.

Source: OECD stat; FAOSTAT (2013), *On-line database*. FAO, http://faostat.fao.org; AQUASAT (2013), *On-line database*. FAO, http://www.fao.org/nr/water/aquastat/dbase/index.stm .

Characteristics of agriculture and agri-food sectors

Farming in Australia spans across three broad zones: pastoral, wheat–sheep and high rainfall zones (Figure 2.2). Large parts of the country's landscape comprise the pastoral zone, which is only suited to low-intensity grazing. Much of it is characterised by low rainfall, less fertile soils and large areas of beef and sheep farming. The principal farming activities in the wheat–sheep zone are winter cropping and livestock grazing. Most of Australia's sheep flock is in this zone. Prime lamb and beef production are undertaken in the high rainfall zone. Much of the dairy industry is found in the coastal areas, along with some dairying in inland irrigated areas. A number of agricultural industries depend on irrigation. These have developed around irrigation schemes, particularly in the Murray–Darling Basin (which covers parts of the mainland eastern states as well as South Australia).

Broadacre farms, essentially non-irrigated crops, cattle and sheep, contribute 54% to the gross value of agricultural production and make up around 53% of agricultural businesses. High value horticultural industries also contribute significantly to the gross value of agricultural production, accounting for 16% in 2011/12 (ABS 2012; ABARES data).

In 2012/13, Australia's agricultural output exceeded nearly by half its volume in 1990/91 (Figure 2.3), however output variations during this period were considerable. Most producers rely on seasonal rainfall to support crop and pasture growth, but Australia has one of the most variable climates in the world. As a result, production, particularly cropping, fluctuates strongly. In drought years, agricultural production can decrease sharply, as was the case following the big droughts in 1994/95, 2002/03 and 2006/7 when total output fell by almost 10% or more.

Australian agriculture has a strong export focus. Around 60% of the gross value of farm production is typically exported, although the share was closer to 75% in 2011/12 (ABARES, 2012). Agri-food exports reached nearly AUD 38 billion in 2012/13. Australia is the second largest world exporter of sheep meat and wool, and is among world's top suppliers of wheat, raw and refined sugar and cotton lint.

Figure 2.2. Australia's broadacre zones

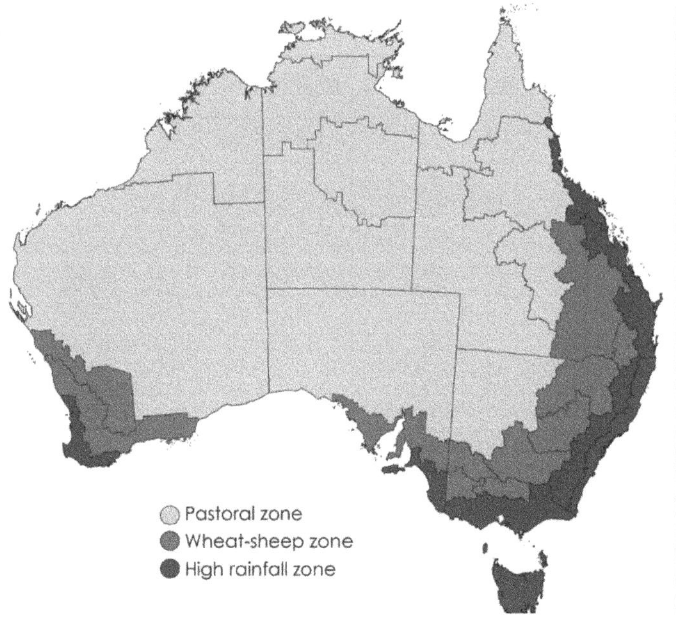

Source: ABARES.

Figure 2.3. Australia's agricultural output indices, 1990-2012

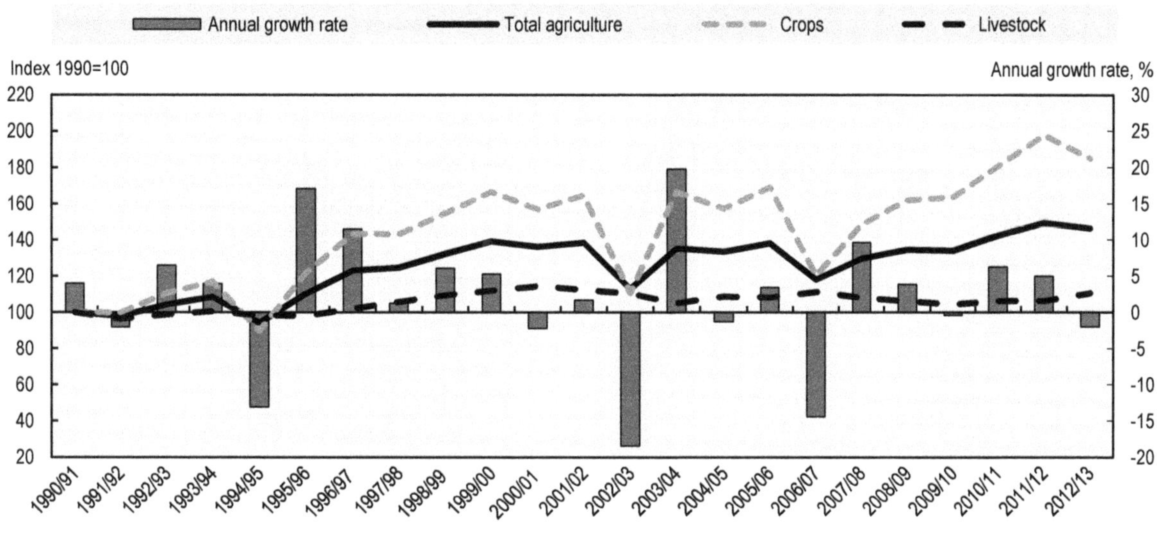

Source: ABARES (2013), *Agricultural Commodity Statistics*, Australian Bureau of Agricultural and Resource Economics and Sciences, Canberra, http://daff.gov.au/abares/publications.

StatLink http://dx.doi.org/10.1787/888933249681

Figure 2.4. Australia's agro-food trade, 1995-2012

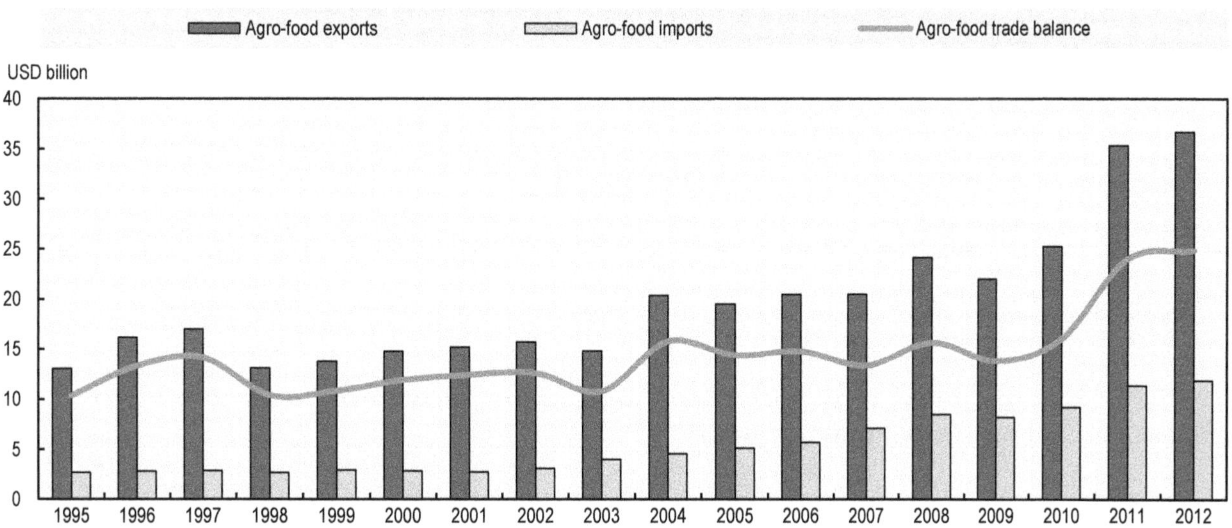

Source: UN Comtrade (2013), On-line database, UN. http://comtrade.un.org/.

StatLink http://dx.doi.org/10.1787/888933249695

Asia is a growing destination for Australia's agricultural production. Asian markets accounted for over 60% of the value of agricultural exports in 2011/12 (ABARES, 2012). The main destinations were Japan, China, Indonesia and Korea. At the same time, exports to Europe have generally declined and exports to the United States have increased, each accounting for around 10% of the value of agricultural exports.

Agriculture's export orientation leads to a relatively high exposure of the Australian economy to agricultural and agro-industry trade (Figure 2.5). It is, however, less significant than for other large agricultural exporters, such as New Zealand, Chile, Indonesia, and Brazil.

Figure 2.5. Australia's exposure to trade in agriculture and food products in international comparison

A. Agro-food exports and imports in % of GDP
2010-12 average

B. Agro-food exports and imports of value-added
as a % of agro-food adjusted GDP (value-added)[1, 2], 2009

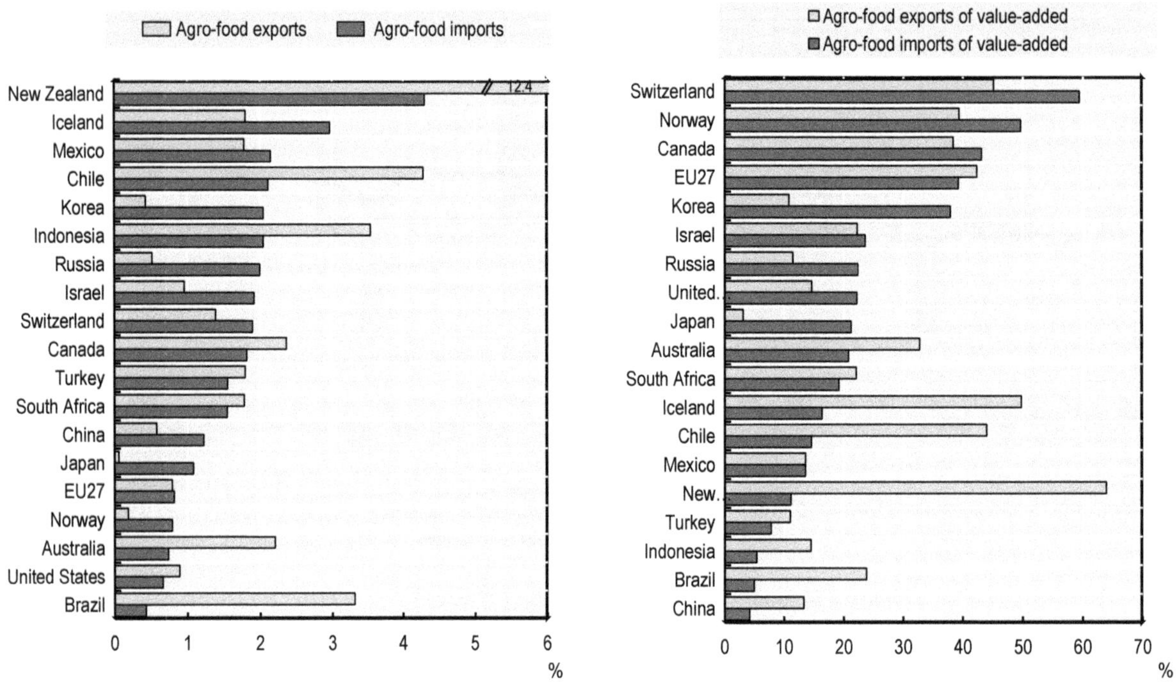

1. Value-Added embodied in Foreign Final Domestic Demand shows how industries export value both through direct final exports and via indirect exports of intermediates through other countries to foreign final consumers. They reflect how industries (upstream in a value-chain) are connected to consumers in other countries, even where no direct trade relationship exists. The indicator illustrates therefore the full upstream impact of final demand in foreign markets to domestic output. It can most readily be interpreted as 'exports of value-added'.

2. Foreign Value-Added embodied in Final Domestic Demand shows how industries abroad (upstream in a value-chain) are connected to consumers at home, even where no direct trade relationship exists. It can most readily be interpreted as 'imports of value-added'.

Source: International Trade by Commodity Statistics (ITCS) Database and OECD-WTO Trade in Value-Added Database, 2013. http://www.oecd.org/industry/ind/measuringtradeinvalue-addedanoecd-wtojointinitiative.htm.

StatLink http://dx.doi.org/10.1787/888933249704

Productivity and sustainability performance

Trends in agricultural productivity

Productivity growth has been central to the continued viability and competitiveness of Australian farm businesses. Over time, productivity growth has helped to maintain farm profitability in the face of a declining trend in the terms of trade (output prices relative to input prices) and has driven agricultural output growth in Australia.

The estimates by the Australian Bureau of Agricultural and Resource Economics and Sciences (ABARES) of agricultural productivity show that since the late 1970s, total factor productivity (TFP) has driven output growth in the broadacre and dairy industries, with declining input use in both industries. Between 1977/78 and 2011/12, TFP in broadacre agriculture averaged around 1% a year. Over this period, the broadacre industry maintained output levels (output growth close to zero) despite reduced input use (–0.9% a year). Dairy productivity growth averaged 1.6% a year. As with broadacre agriculture, productivity gains drove growth in total outputs (1.3% a year on average), while input use declined (–0.3% a year on average) (Figure 2.6).

Figure 2.6. Annual input, output and TFP growth rates in the broadacre and dairy industries between 1977/78 and 2011/12

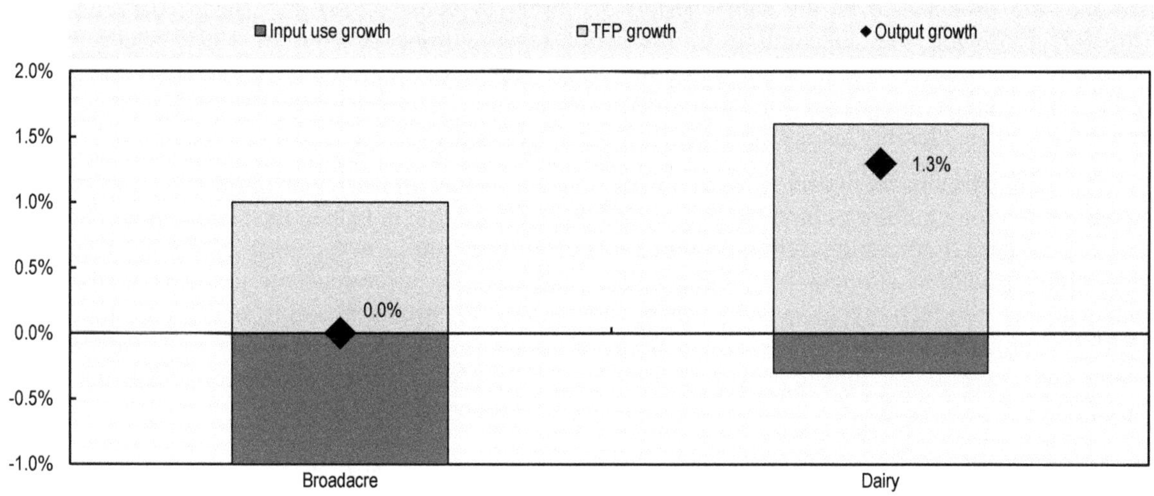

Source: Gray et al. (2014), "Productivity in the Broadacre and Dairy Industries", in Agricultural Commodities: March Quarter 2014, Australian Bureau of Agricultural and Resource Economics and Sciences, Canberra, http://daff.gov.au/abares/publications.

StatLink http://dx.doi.org/10.1787/888933249718

Figure 2.7. Annual average TFP productivity growth in broadacre and dairy industries, by period

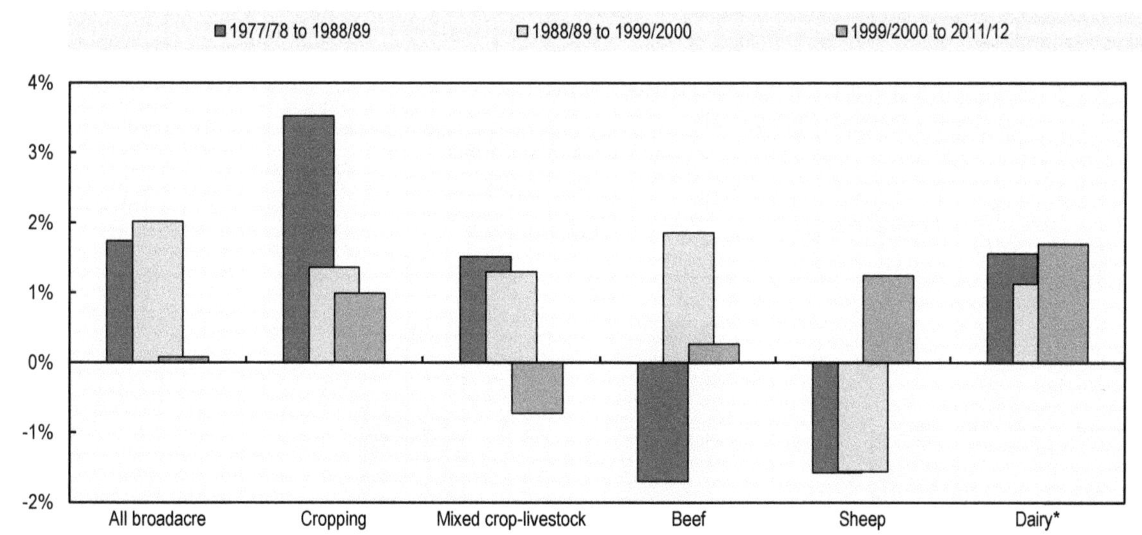

* Dairy total factor productivity series commenced in 1979/80.

Source: Gray *et al.* (2014), "Productivity in the Broadacre and Dairy Industries", in Agricultural Commodities: March Quarter 2014, Australian Bureau of Agricultural and Resource Economics and Sciences, Canberra, http://daff.gov.au/abares/publications.

StatLink http://dx.doi.org/10.1787/888933249725

Industry aggregate TFP trends disguise marked differences across specific sectors (Figure 2.7). For cropping specialists, the TFP growth rate averaged 1.6% a year between 1977/78 and 2011/12, higher than the rate observed over the same period on farms in beef (0.8%) and sheep (0.1%) farming. However, following the dismantling of the wool reserve price scheme in 1991, sheep industry productivity has increased at an average rate of 1.4% a year since the mid-1990s. The dairy industry has realised an average annual productivity growth of around 1.6% since the late 1970s. Productivity growth varies considerably across farms and regions (Dahl et al., 2013).

Productivity gains were driven by advances in technologies, in particular in the cropping and dairy industries. A strong culture of innovation amongst Australian farmers helped the diffusion and uptake of new technologies. Economic reforms provided a significant stimulus to technical change as a source of productivity growth by strengthening the incentives and capacity of producers to innovate. Reforms also contributed to productivity growth by easing constraints to structural adjustment. The analysis by ABARES shows that structural adjustment (more efficient resource use across farms, as well as farms exiting and entering the industry) was an important driver of TFP growth in broadcare agriculture (Box 1.1).

Box 1.1. Structural adjustment as a driver of TFP growth in Australian agriculture

Agricultural productivity analysis by ABARES provides an insight into the significance of resource reallocation effects – gains from the reallocation of resources due to farms exiting and entering broadacre agriculture, and due to resources moving between farms with differing levels of productivity – as a driver of industry-level productivity growth.

Based on the Olley and Pakes (1996) method, annual industry-level broadacre productivity growth was decomposed into average on-farm productivity growth and resource reallocation effects (Figure 2.8).

Although resource reallocation slightly detracted from on-farm productivity gains between 1977/78 and 1989/90 as farmers pursued higher profits in less efficient industries, it has subsequently played a more important role. Efficiency gains from the reallocation of resources accounted for over a third of broadacre TFP growth between 1989/90 and 1999/2000 and partly offset the effects of declining on-farm productivity between 1999/2000 and 2009/10. A key factor affecting broadcare productivity performance between 1989/90 and 1999/2000 was the demise of the Wool Reserve Price Scheme in 1991. This led to significant changes in the structure of Australia's broadacre industries, as many farmers left the wool industry for crop and sheep meat production.

Figure 2.8. Contribution of on-farm productivity growth and resource re-allocation to TFP growth in the broadacre industry

Percentage per year

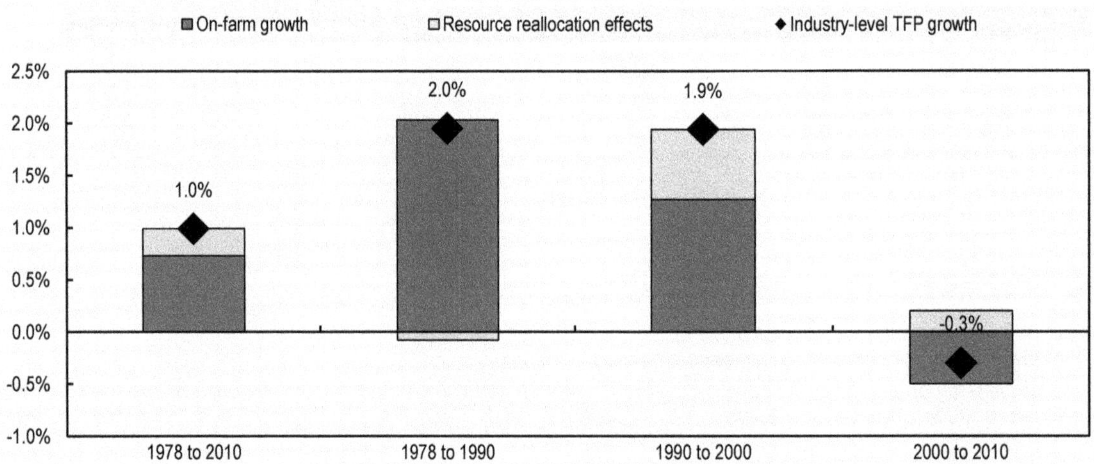

StatLink http://dx.doi.org/10.1787/888933249731

The decomposition is based on Olley-Pakes method: productivity growth due to resource re-allocation effects is measured by the co-variance between changing weights and farm-level productivity growth. Years correspond to financial years.

Source: Gray, E.M. et al. (2014), *Australian Agricultural Productivity Growth: Past Reforms and Future Opportunities*, http://www.daff.gov.au/ABARES/Pages/publications/default.aspx.

Notwithstanding the long period of productivity growth, it considerably slowed down in the 2000s (Figure 2.7 and Figure 2.8). Although the dairy and sheep sectors realised significant productivity improvements, most of the broadacre segments experienced either a significant slow-down of productivity growth or negative rates.

A slow-down of Australia's agricultural productivity is also revealed by the cross-country productivity study by the USDA, although a cross-country comparison of productivity growth rates needs to be made with care (USDA, 2013).[2] Australia showed one of the fastest agricultural productivity improvements in the 1990s (Figure 2.9). As concluded by Nossal and Sheng (2013), this enabled the country to maintain competitiveness *vis-à-vis* the United States and to improve relative to Canada. Nevertheless, Australia saw a deceleration of productivity growth rates in the following decade which was by far more significant than in these two countries and some other OECD economies also experiencing diminishing productivity growth rates (e.g. France, Mexico, Korea, and New Zealand).

Figure 2.9. Average annual growth rates in agricultural TFP, international comparison

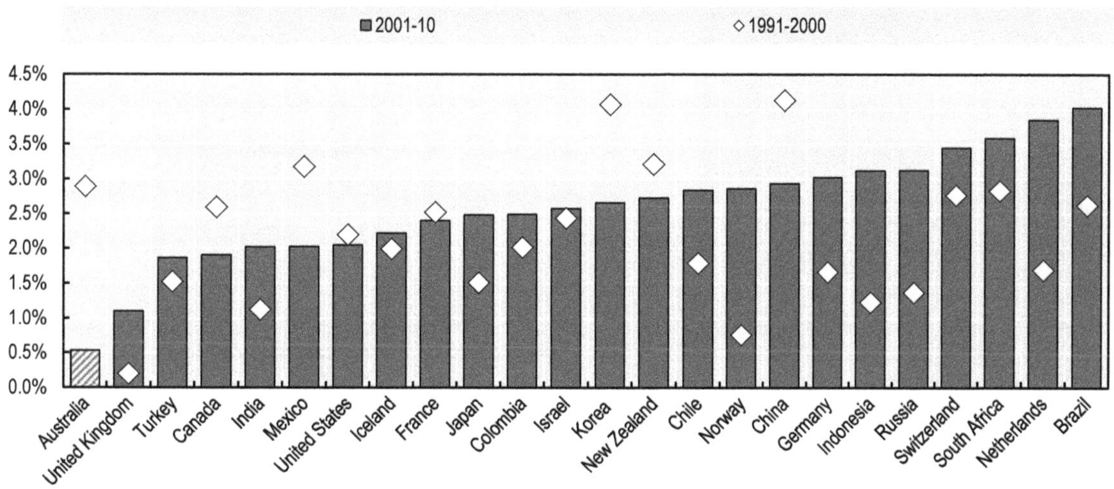

Source: USDA (2013), *Economic Research Service International Agricultural Productivity database*. www.ers.usda.gov/data-products/international-agricultural-productivity/documentation-and-methods.aspx#excel.

StatLink http://dx.doi.org/10.1787/888933249747

An analysis by ABARES points to a structural break in the agricultural productivity trend in Australia occurring in the mid-1990s. Several Australian studies highlight climatic factors that have contributed to this effect, namely the fact that Australia experienced poor seasonal conditions over much of the 2000s. For example, drought is estimated to have reduced the output of mixed crop–livestock producers and cropping specialists by around 11% since 2000 (Hughes et al., 2011). Based on econometric analysis, Sheng et al. (2011) found that climate conditions were indeed a significant factor affecting the TPF trend, although it did not fully explain it.

There is evidence that the rate of technical change in some key agriculture industries is slowing down and that the productivity gap between the most productive and average farm is widening. Agriculture sector productivity is driven by the changes in farm-level productivity and the shifts in farm scale and input mix. Farm-level productivity increases as a result of technical change (i.e. expansion of production frontier due to the best farms becoming more productive) and through a technical efficiency change (i.e. other farms moving closer to the production frontier). ABARES analysed the rates of production frontier and technical efficiency changes based on climate-adjusted agricultural productivity time series. This analysis revealed that since the late 1970s, technical change has been the primary driver of long-run productivity growth, although its rate slowed in the 2000s, whereas technical efficiency declined over the whole period of analysis (Figure 2.10). Thus, since the beginning of the

2000s, the best farms in Australia have been improving their productivity more slowly, while the rest of the farms were not catching up in productivity with the best performers. Both trends have been a drag on productivity growth, which was partly offset by the impact of farm scale and input mix adjustment in response to prevailing input and output prices.

Figure 2.10. Average annual growth of climate-adjusted TFP in broadcare cropping industry: Contribution of changes in farm-level productivity and farm scale and input mix

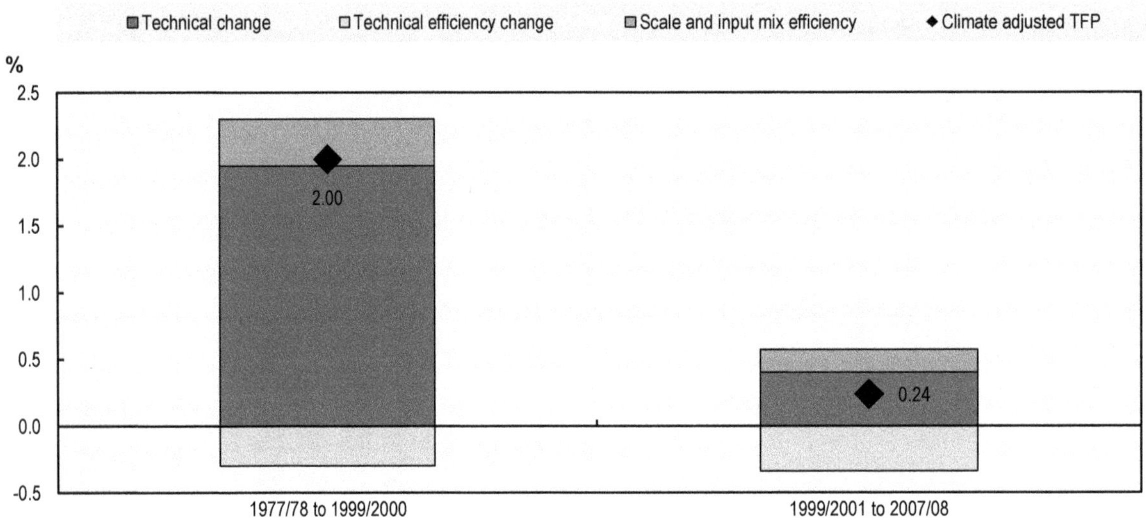

Source: Hughes et al (2011), *Productivity Pathways: Climate Adjusted Production Frontiers for the Australian Broadacre Cropping Industry,* Research report No 11.5, Australian Bureau of Agricultural and Resource Economics and Sciences, Canberra, http://daff.gov.au/abares/publications.

StatLink http://dx.doi.org/10.1787/888933249750

One insight into possible causes of the declining rate of technical change and the widening productivity gap between the best and average performers comes from the analysis of the effects of the "usable knowledge stock" available to farmers. It is proxied by past public investments in agricultural R&D accumulated over a long period.[3] In fact, Australia has experienced stagnating real public investments in agricultural R&D since the late 1970s. When the joint effects of climate and R&D investment were accounted for, the structural break in the productivity trend was no longer apparent.

More broadly, the speed with which technical change is occurring depends on farmers having the incentives, capacity, and willingness to adopt innovations. Many inter-related factors determine a farmer's decisions. Firstly, the innovation process depends on the farmer's personal characteristics, such as risk aversion, attitude to learning, level of motivation, or past experience with innovation. Other types of factors include the specific characteristics of farm business, such as the farmer's age, education level, business experience, income level, and farm size. There is the overall business environment beyond the individual farm enterprise which generates economic incentives or disincentives for businesses to innovate. The business environment is in turn influenced by government policies, e.g. the degree to which they ensure macro-economic stability, allow for competition and flexibility of input and factor markets and adequately provide public goods. Policies may also influence farmers' individual characteristics, such as risk aversion by equipping farmers with the instruments to manage risks, or the level of education by establishing a performing educational system. The focus of this review is to assess the overall business environment from the perspective of business conditions to invest and innovate and how government policies affect these conditions.

Sustainability performance of agriculture

Australia made considerable efforts through national, state and community programmes and private initiatives, to improve environmental sustainability of the agriculture sector and increase its preparedness to potential impacts of climate change. Progress has been achieved in many areas, in particular in water and land management. However, environmental performance of Australian agriculture remains mixed and its environmental footprint is still substantial.

The share of total land in agriculture fell markedly in Australia during the past two and a half decades, this trend establishing amid considerable fluctuations of agricultural output during this period (Table 2.3). Extensive pastoral farming dominates the land use pattern, but there has been some shift to more intensive land uses for permanent and arable crops. The past overgrazing and land clearing create pressures on the environment. One million hectares were cleared in Australia over the decade of the 2000s, but a balance between land clearing and regrowth has been achieved by 2010. Still, the vegetation loss leading to land erosion remains substantial. Studies indicate that about three-quarters of all soil losses occur on extensively grazed native pasture, while the rate of erosion is the highest in the zones of intensive land use (in the mid-2000s, around 70% of cropping areas had erosion rates by far exceeding the average natural rate of erosion). Wind and water erosion hazards have reduced over time due to changed cropping and tillage management practices, but continued improvement in groundcover management is needed to minimise erosion risk under a drying climate.

The 2011 State of the Environment report identified widespread soil acidification in southern Australia, notably in the Western Australian wheat belt and in intensive land use systems, such as tropical horticulture, cane and dairying. Acidification is exacerbated by farming practices including the use of acidifying fertilisers, and is a major constraint to biomass production and soil carbon storage. Increased understanding of this issue has led to substantial increases in rates of soil testing and lime use over the past decade, but they still fall short of what is needed to treat existing and ongoing acidity. Salinisation of soils due to changes in the hydrological cycle wrought by land clearing and irrigation has been an important issue, given Australia's geological and climatic conditions favouring salt concentration in the landscape. Drying of the climate over southern Australia has reduced concerns about salinity in the past decade.

The dominance of extensive agricultural systems leads to fertiliser and agrochemical footprint in Australia being relatively small compared to the countries with more intensive agriculture. The available data on per hectare consumption of fertilisers and pesticides indicates that compared to 1990, the intensity of agrochemicals use increased. Livestock operations represent another source of nutrient pollution, although livestock density in Australia has remained relatively stable over time (Table 2.3). OECD nutrient balances compare the input and uptake of nutrients through agricultural activity.[4] The nitrogen balance for Australia shows that it has one of the lowest nitrogen surpluses per hectare among OECD countries – in 2007-09 it was less than 14 kg per hectare in nitrogen equivalent compared to 63 kg in OECD on average, 23 kg in Canada, or 49 kg in New Zealand (Figure 2.A1.1). In the Australian phosphorous balance, the input and uptake of the nutrient were marginally positively balanced throughout the 1990s, with the balance turning to zero in 2007-09 (Figure 2.A1.1). Australia followed the tendency of diminishing nutrient surpluses observed in the majority of OECD countries. A comparison of the changes in Australia's nitrogen and phosphorous balances with the changes in agricultural output shows that in the 1990s the expansion of production occurred at a decreased nitrogen surplus and although the phosphorous surplus increased, this occurred less rapidly than the growth in output. In the 2000s, agricultural output was falling, as did nutrient surpluses, but the latter decreased at much faster rates (Figure 2.11). These trends suggest an improvement in environmental sustainability of agricultural chemical nutrient use over the recent decade and a reduction of respective risks of soil, air, and water pollution.

Table 2.3. Australia's selected agro-environmental indicators, 1990-2011

	1990-92 average	1998-2000 average	2008-10 average	Annual % change over the period	Annual % change over the period
				1990-92 to 1998-2000	1998-2000 to 2008-10
Land					
Agricultural land (% of land area)	60.4	59.6	53.1	-0.2	-1.1
Agricultural land use (in %)					
Share of arable in agricultural land area, % [a]	10.1	10.0	10.9	-0.2	0.9
Share of permanent cropland in agricultural land area, % [b]	0.0	0.1	0.1	3.5	6.9
Share of permanent pasture in agricultural land area, % [c]	89.9	90.0	89.0	0.0	-0.1
Conservation area over 30% ground cover (% of agricultural land) [d]		2.0	4.1		26.3
Organic area (% of agricultural land) [e]	..	2.5	2.9	..	2.5
Water					
Total agricultural freshwater withdrawals, million m[3f]	14,498	15,033	7,359	0.6	-7.6
Share of agriculture in total withdrawals, % [g]	79.1	65.2	52.2	-1.7	-4.4
Irrigated area [h]	2,057	2,476	1,817	3.8	-4.3
Irrigation water application rates [i]	8.7	4.9	3.6	-13.2	-4.0
Air and climate change					
Agricultural GHG emissions					
1 000 metric tons of CO_2 equivalent	86,151	89,870	83,370	0.5	-0.7
% of total national emissions	20.5	18.6	15.2	-1.2	-2.0
Livestock					
Livestock number per ha of agricultural area [1 j]	0.5	0.5	0.5	-0.3	-0.7
Fertiliser and pesticide use					
Fertiliser use on arable and permanent crop area, tonnes per 1 000 ha [2 k]	24.3	45.1	39.9	8.1	-1.2
Pesticide use on arable and permanent crop area, tonnes per 1 000 ha [3 l]	0.4	0.8	0.7	8.1	-0.7
Nitrogen balance per hectare, kg [m]	16.2	16.1	13.7	-0.1	-1.7
Phosphorous balance per hectare, kg [n]	0.6	0.8	0.0	2.3	-27.5
Agricultural ammonia emissions					
1 000 metric tons	..	..	..	..	..
% of total emissions	..	..	..	..	..
Energy					
Energy use in agriculture and forestry (% of total energy use) [q]	2.3	2.3	2.7	-0.1	1.9
Bioenergy production (% of total renewable energy production) [r]	72.3	76.7	78.2	0.7	0.2

.. Not available; nc: not calculated

1. Cattle, buffaloes, pigs, sheep, goats, and poultry.

2. In nitrogen and phosphate nutrients.

3. In active ingredient.

d: Data for 1998-2000 average is 2005; data for 2008-10 average is 2008.
e: Data for 1998-2000 average is 2000-02.
f: Data for 1990-92 is 1994-96 average, value for 1990-2002 is 2001; and value for 2008-10 is 2010.
g: Data for 1990-92 average is 1994; data for 1998-2000 average is 2005; data for 2008-10 is 2010.
h: Data for 1990-92 average is 1997; data for 1998-2000 average is 2001-03.
i: Data for 1990-92 average is 1997; data for 1998-2000 average is 2001.
l: Data for 2008-10 average is 2004-06.
m,n,q,r: Data for 2008-10 average is 2007-09.

Source: Data a,*b,c,d,j,k,l,q,r* retreived from FAO Agri-Environmental Indicators, 2014 from http://faostat.fao.org on 23 June 2014. All other data from *OECD/Eurostat Agri-Environmental Indicator Database*, http://epp.eurostat.ec.europa.eu.

Figure 2.11. Average annual changes in nutrient balances and production volumes

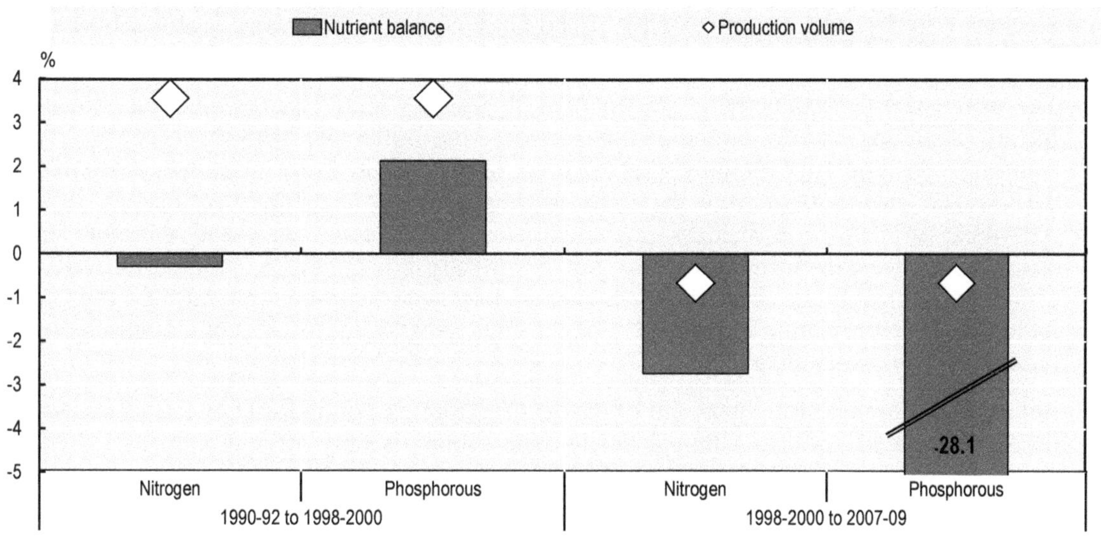

Source: *OECD/Eurostat Agri-Environmental Indicator Database*, http://epp.eurostat.ec.europa.eu; FAOSTAT, 2013.

StatLink http://dx.doi.org/10.1787/888933249768

Water is Australia's scarce agricultural resource. The agricultural sector is the major water consumer accounting for 52% of total freshwater withdrawals in 2008-10 (Table 2.3), with approximately 90% of that share representing water used for irrigation. Irrigated area is small, occupying only around 0.4% of total agricultural land (concentrated in the Murray-Darling Basin), but production on these lands contributes a disproportionately high share to the sector's overall output. Irrigation is used in virtually all agricultural industries, including for bringing water to pastures. Australia took major steps to improve water management, notably, through important reform of water use rights and development of water markets. The trends in water use in the 2000s suggest a declining pressure from agriculture on this resource. Water use in agriculture during this decade decreased by 7.6% per year compared to a 5.2% reduction in total national water withdrawals; this reflects the shrinkage of irrigated area and substantial gains in efficiency of water use, with water application rates strongly falling in the 1990s and the efficiency improvements continuing in the 2000s. Despite the progress in water use efficiency, water quality remains an important issue (OECD, 2012b; OECD 2013). However, given the insufficient monitoring the impact of agriculture is difficult to assess. There is a problem of water salinity. Coastal regions downstream of large agricultural areas suffer from sediment and nutrient loading, with recent research indicating for example, that agriculture is a key contributor to water quality issues in Australia's Great Barrier Reef – this aqua-system has been subject to successive water quality protection plans (OECD, 2013). The Australian Government and state and territory governments co-operate on a national approach to water quality management outlined in the National Water Quality Management Strategy. This strategy includes a range of guiding documents that set out ways to improve water quality and manage risks, and while it is non-mandatory, it is utilised by all state and territory governments for managing the quality and supply of water that is fit for purpose, including those fit for agricultural purposes.

Notes

1. Sections 1.3, 1.4 and 1.5 draw extensively on Gray et al. (2014a), Gray et al. (2014b) and Dahl et al. (2013).

2. The differences in productivity growth rates depend on many factors, such as climatic factors prevailing during particular time periods, the initial productivity levels and the countries' position in exploiting the catch-up opportunities. Note also that the USDA derives TFP growth rates from the sector-aggregate output and input use indexes, while ABARES estimates use indexes constructed from farm-level data and therefore the estimates by the two institutions may not be directly compared. In addition, ABARES estimates cover only the broadacre and dairy sectors, with other agricultural industries not included and representing around 45% of the total gross value of Australian agricultural production (five years ending 2012/13).

3. For any given time, a weighted sum of past expenditure on public R&D over 35 years was used (Sheng et al., 2011). Although R&D expenditures is only an approximation of the usable knowledge stock and do not reflect the degree to which this stock is taken up, the findings of this study suggest a link between the declining dynamics of agricultural productivity in Australia and the performance of the national knowledge system.

4. Gross nutrient balances (N and P) are calculated as the difference between the total quantity of nutrient inputs entering an agricultural system (mainly fertilisers, livestock manure), and the quantity of nutrient outputs leaving the system (mainly uptake of nutrients by crops and grassland. This calculation can be used as a proxy to reveal the status of environmental pressures, such as declining soil fertility in the case of a nutrient deficit, or for a nutrient surplus the risk of polluting soil, water and air (OECD, 2013c).

References

ABARES (2013), *Agricultural Commodity Statistics 2013*, Australian Bureau of Agricultural and Resource Economics and Sciences, Canberra. daff.gov.au/abares/publications.

ABARES (2012), *Agricultural Commodity Statistics 2012*, Australian Bureau of Agricultural and Resource Economics and Sciences, Canberra. daff.gov.au/abares/publications.

ABS (2012), *Australian Social Trends*, Cat. No. 4102.0, Australian Bureau of Statistics, Canberra.

Australian Bureau of Meteorology (2014), *State of Climate 2014*, Canberra, http://www.bom.gov.au/state-of-the-climate/.

Credit Suisse (2013), Global Wealth Report 2013, Research Institute Credit Suisse https://publications.credit-suisse.com/tasks/render/file/?fileID=BCDB1364-A105-0560-1332EC9100FF5C83.

Dahl, A., Leith, R. and Gray, E. M. (2013), "Productivity in the broadacre and dairy industries", *Agricultural Commodities*, March Quarter, Australian Bureau of Agricultural and Resource Economics and Sciences, Canberra, pp. 200–20, http://www.daff.gov.au/abares/publications

DFAT (2012), *Feeding the Future : a Joint Australia–China Report on Strengthening Agricultural Investment and Technological Cooperation to Improve Food Security*, Australian Department of Foreign Affairs and Trade (Australia) and Ministry of Commerce of China, December. http://dfat.gov.au/about-us/publications/trade-investment/feeding-future/Documents/feeding-the-future.pdf. .

Gray, E.M., Leith, R. and Davidson, A. (2014a), "Productivity in the Broadacre and Dairy Industries", *Agricultural Commodities*, March Quarter, Australian Bureau of Agricultural and Resource Economics and Sciences, Canberra http://www.daff.gov.au/abares/publications.

Gray, E.M., Oss-Emer, M. and Sheng, Y. (2014b), *Australian Agricultural Productivity Growth: Past Reforms and Future Opportunities*, Australian Bureau of Agricultural and Resource Economics and Sciences Research Report 14.2, Canberra, February, http://www.daff.gov.au/abares/publications.

Hughes, N., Lawson, K., Davidson, A., Jackson, T. and Sheng, Y. (2011), *Productivity Pathways: Climate Adjusted Production Frontiers for the Australian Broadacre Cropping Industry*, Research report No 11.5, Australian Bureau of Agricultural and Resource Economics and Sciences, Canberra, http://www.daff.gov.au/abares/publications

Nossal, K. and Sheng, Y. (2013), *Cross-country Comparisons of Agricultural Productivity: an Australian Perspective*, RIRDC publication No. 13/011, Rural Industries Research and Development Corporation, Canberra.

OECD (2014), *OECD (2014), Society at a Glance 2014: OECD Social Indicators,* OECD Publishing, Paris.
DOI: http://dx.doi.org/10.1787/soc_glance-2014-en . OECD Social Indicators on-line http://www.oecd.org/els/soc/sag2014-national.htm.

OECD (2013), *OECD Compendium of Agri-environmental Indicators*, OECD Publishing, Paris, DOI: http://dx.doi.org/10.1787/9789264181151-en.

OECD (2012a), *OECD Economic Surveys: Australia 2012*, OECD Publishing, Paris.
DOI: http://dx.doi.org/10.1787/eco_surveys-aus-2012-en

OECD (2012b), *Water Quality and Agriculture: Meeting the Policy Challenge*, OECD Studies on Water, OECD Publishing, Paris.
DOI: http://dx.doi.org/10.1787/9789264168060-en

Olley, S and Pakes, A (1996), 'The dynamics of productivity in the telecommunications industry', Econometrica, vol. 64, no. 6, pp. 1263–98.

Penm, J. and Glyde, P. (2007), "Australian Agriculture—Key Issues for the Future", *Australian Commodities*, March, Australian Bureau of Agricultural and Resource Economics, pp. 22–6.

Sheng, Y, J.D. Mullen, and S. Zhao (2011), A *turning point in agricultural productivity: consideration of the causes*, research report 11.4 for the Grains Research and Research and Development Corporation, Australian Bureau of Agricultural and Resource Economics and Sciences, Canberra, May 2011, http://www.daff.gov.au/abares/publications .

UN Comtrade (2013), On-line database, UN. http://comtrade.un.org/.

USDA (2013), Economic Research Service International Agricultural Productivity database. www.ers.usda.gov/data-products/international-agricultural-productivity/documentation-and-methods.aspx#excel.

WDI (2014), World Development Indicators Database. World Bank, http://data.worldbank.org/data-catalog/world-development-indicators.

Annex 2.A1

Australia's nitrogen and phosporous balances

Figure 2.A1.1. Nitrogen balance per hectare of agricultural land, OECD countries, 1990-2009

Kg of nitrogen per hectare

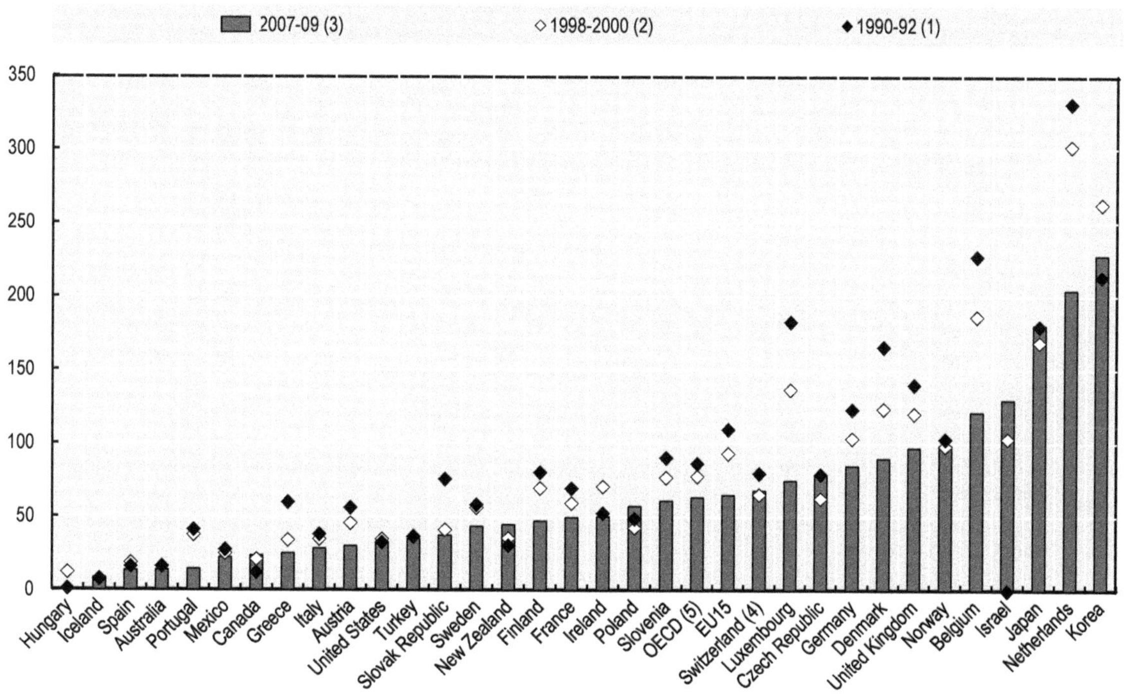

1. Data for 1990-92 average refer to the year 1990 for the United Kingdom; the 1992-94 average for Slovenia; and the 1995-97 average for Portugal.

2. Data for 1998-2000 average refer to the year 2000 for the United Kingdom; and the 2000-02 average for Israel and Portugal.

3. Data for 2007-09 average refer to the 2006-08 average for Belgium, France, Greece, Hungary, Italy, Luxembourg, Mexico, Netherlands, Slovenia and Switzerland.

4. In the case of Switzerland, total agricultural area includes summer grazing.

5. The OECD total excludes Chile, Estonia and Israel.

Source: *OECD/Eurostat Agri-Environmental Indicator Database*, http://epp.eurostat.ec.europa.eu; and national data for Spain.

StatLink http://dx.doi.org/10.1787/888933249773

Figure 2.A1.2. Prosphorus balance per hectare of agricultural land, OECD countries, 1990-2009

Kg of phosphorus per hectare

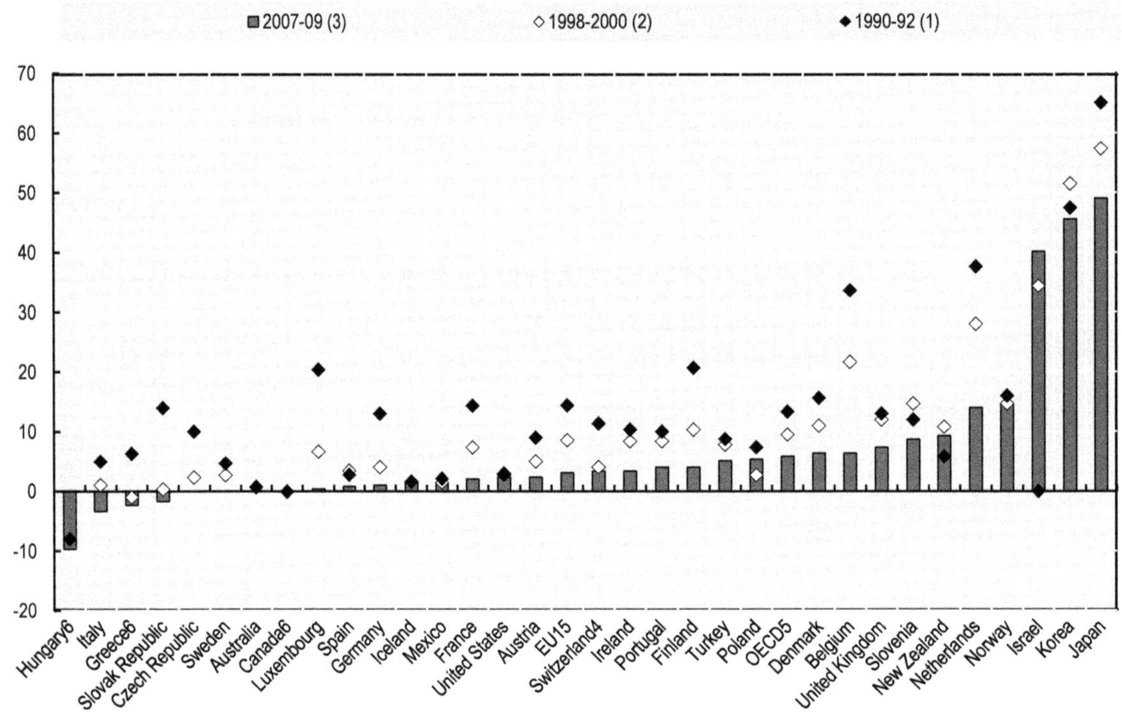

1. Data for 1990-92 average refer to the year 1990 for the United Kingdom; and the 1995-97 average for Portugal and Slovenia.

2. Data for 1998-2000 average refer to the year 2000 for the United Kingdom; and the 2000-02 average for Israel, Portugal and Slovenia.

3. Data for 2007-09 average refer to the 2006-08 average for Belgium, Estonia, France, Germany, Greece, Hungary, Italy, Luxembourg, Mexico, Netherlands, Slovenia and Switzerland.

4. In the case of Switzerland, total agricultural area includes summer grazing.

5. The OECD total excludes Chile, Estonia and Israel.

6. For Canada, Greece and Hungary, the average annual percentage change refers to change in phosphorus deficit.

Source: *OECD/Eurostat Agri-Environmental Indicator Database*, http://epp.eurostat.ec.europa.eu; and national data for Spain

StatLink http://dx.doi.org/10.1787/888933249782

Chapter 3

Economic stability and trust in institutions in Australia

This chapter outlines the importance of economic stability and public institutions in fostering public and private investment. It provides an overview of the performance of the overall economy, outlines macroeconomic developments and challenges, explains the federal-provincial governance system, and presents an evaluation of public institutions.

The statistical data for Israel are supplied by and under the responsibility of the relevant Israeli authorities. The use of such data by the OECD is without prejudice to the status of the Golan Heights, East Jerusalem and Israeli settlements in the West Bank under the terms of international law.

Innovation is an activity with a medium- to long-term horizon that requires a stable environment. A stable macroeconomic framework and good public governance and institutions are a prerequisite for unleashing and fostering innovation. Under conditions of economic and political instability and weak institutions, potential innovators will perceive significant risks to longer term undertakings and are unlikely to invest in activities that do not generate rapid returns (OECD, 2010).

Macroeconomic policy environment

The dynamic development of the Australian economy was largely a response to economic reforms undertaken since the early 1980s. These reforms were broad-ranging and concerned all areas of economic activity, including trade, FDI, taxation, and in the finance and labour markets (Annex 2.A1). This success is also due to the government's exemplary efforts to establish a regulatory environment promoting competition and productivity growth.

Australia managed the macro-economic effects of the recent commodity boom better than during similar episodes in the past. The robust macroeconomic framework with the flexible exchange rate mechanism, and flexible labour and competitive product markets allowed inflation and unemployment to be contained and stable growth to be maintained (Table 3.1) (OECD, 2012a).

Table 3.1. Australia's key indicators of macroeconomic policy

	2001	2005	2006	2007	2008	2009	2010	2011	2012	2013	2014 e	2015 e	2016 e
Real GDP growth, %	2.7	3.3	2.6	4.7	2.5	1.6	2.3	2.6	3.6	2.4	3.1	2.5	3.0
General government financial balance[1]	0.1	1.9	2.2	1.5	-0.6	-4.7	-5.1	-3.6	-3.0	-1.3	-3.3	-2.0	-1.5
General government gross debt[2]	19.0	12.6	12.1	12.3	14.0	19.3	23.3	28.0	31.8	33.0	36.2	38.1	39.3
Current account balance[1]	-2.1	-5.9	-5.8	-6.7	-4.8	-4.7	-3.6	-3.0	-4.4	-3.3	-3.6	-4.1	-3.6
Exchange rate, AUD per USD[3]	1.94	1.31	1.33	1.20	1.20	1.28	1.09	0.97	0.97	1.04	1.11	1.16	1.16
Inflation, annual %, CPI all items	4.4	2.7	3.5	2.4	4.3	1.8	2.9	3.3	1.8	2.4	2.6	2.3	2.6
Unemployment rate, %[4]	6.7	5.0	4.8	4.4	4.2	5.6	5.2	5.1	5.2	5.7	6.1	6.2	5.9

e: *OECD Economic Outlook* estimate.

1. As a percentage of GDP.

2. As a percentage of GDP at market value.

3. End fiscal year; data for 2014 and 2015 is the estimate by the Economic Intelligence Unit.

4. Period average.

5. End year, as a percentage of total labour force.

Source: OECD *Economic Outlook 1996 database* – November 2014. http://dotstat.oecd.org/.

The effects of the 2008-09 global financial crisis on the Australian economy were relatively mild. Australia was one of the few OECD countries to maintain growth in both 2008 and 2009, although it did lose pace. Australia's resilience to the shock was partly due to less exposure to the European markets compared to the Asian region and to prudent economic policies adopted in the years before the crisis. The government's prompt monetary policy response during the crisis, coupled with a large, timely and targeted fiscal stimulus was also a factor. (OECD, 2012a) The World Economic Forum's Global Competitiveness Index ranked Australia in 2013-14 above most OECD countries based on indicators, such as the government's budget balance, gross national savings, inflation, government debt, and the country's credit rating (Figure 3.1). The crisis, nevertheless, was associated with an increase in public debt: since 2008 the government has been running a budget deficit. This has necessitated a tightening of fiscal policy, in particular to avoid undesirable effects on the high sovereign rating and the cost of servicing the public debt. The government's stated objective is to achieve a budget surplus of 1% of GDP by 2023/24. Some of the deficit reduction will happen automatically as growth picks up, but active fiscal measures will also be required. The government's recent proposal for the 2014/15 budget includes spending cuts, tax increases, and an increase in the retirement age (OECD, 2014; EIU, 2014).

The latest *OECD Economic Outlook* estimates Australian growth to remain moderate in 2014 and slow-down in 2015 before gradually accelerating to its potential rate of 3% by 2016. The mining investment boom is set to decline. This puts downward pressure on demand, but may be offset by a gathering momentum in consumption and the export sectors. Macroeconomic policy will need to be focused on helping the economy to rebalance away from investment in the natural resource sector; a budget surplus should be aimed at by the early 2020s in fiscal area (OECD, 2014).

Figure 3.1. Global Competitiveness Index: Index of macroeconomic environment, 2013-14

Scale 1 to 7 (best)

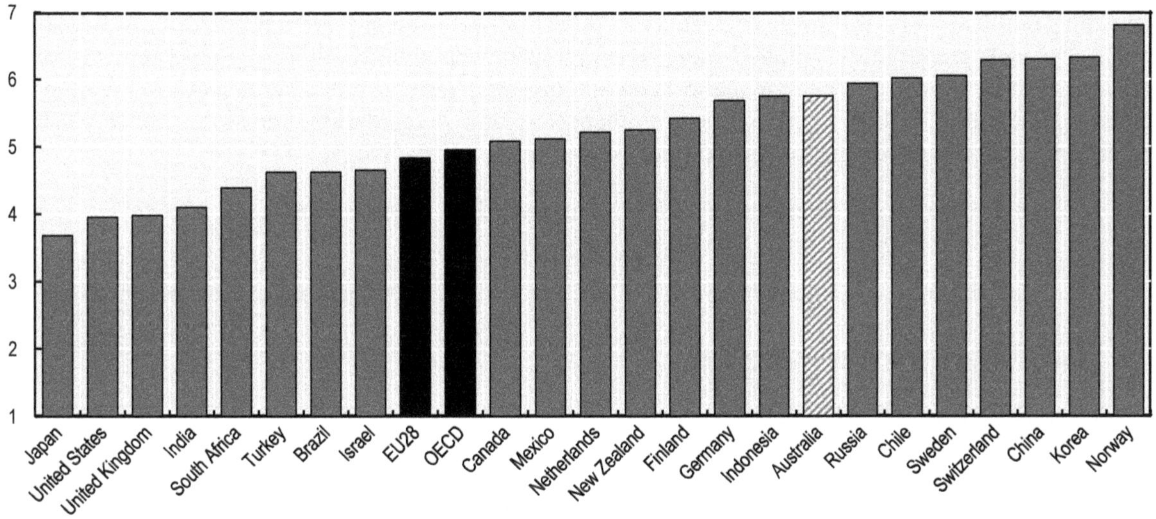

1. Indexes for EU-28 and OECD represent simple averages of member-country indexes.

Source: World Economic Forum (2013), *The Global Competitiveness Report 2013-2014*: Full data Edition, Geneva 2013. http://reports.weforum.org/the-global-competitiveness-report-2013-2014/#=.

StatLink http://dx.doi.org/10.1787/888933249790

Governance and quality of public institutions

Governance systems and institutions fundamentally affect risk perceptions of potential innovators. When government is accountable, transparent, predictable and non-corrupt, institutions work well and economic agents see their risks reduced and are more willing to exploit the opportunities and invest in innovation. Governance systems play an important role in addressing market failure, influencing the behaviour of firms as well as the efficient functioning of farm input and output markets.

Australia is a federal state, with the powers and responsibilities shared by federal, state and local governments. Total government employment is relatively small due to significant privatisation and outsourcing efforts in recent decades. Governance is quite largely decentralised, with a significant weight of sub-national governments in total government employment (over 80%) and general government expenditures (over 45%) (OECD, 2011; OECD, 2012b). Australia has implemented several stages of public management reforms since the mid-1970s resulting in a considerable transformation of the culture and practice of public administration. These changes concerned, for example, fiscal management, the division of responsibilities across all levels of governance, and the establishment of a more transparent and consultative process of policy development and evaluation.

Today, Australia is one of the world's front-runners for the quality of its public governance and institutions. The World Bank's governance indicators consistently position the country in the highest ranks across by the dimensions of governance, such as voice and accountability, political stability, government effectiveness, regulatory quality, rule of law, and control of corruption (WGI, 2013). Australian business perception of the quality of public institutions is more favourable than on average across OECD countries, but Australia still ranks behind some OECD members in government efficiency (Figure 3.2). According to the 2012 Gallup World Poll, almost 42% of Australian citizens had confidence in their national government, above the 40% OECD average share. However, as elsewhere across the OECD, the confidence level of Australians in government has declined since the pre-crisis year of 2007 when it was 53% (OECD, 2013). The current political setting is featured by a change in the ruling political force following the return to power of the Liberal-National coalition in September 2013. The new government initiated extensive policy development activities. Some reforms of the previous Labour Party government have been undone and new policy approaches or broad policy reviews announced in key areas, such as competition and deregulation, taxation, labour relations, education, territorial development, agricultural policy, and climate change.

Figure 3.2. Global Competitiveness Index: Index of quality of public institutions, 2013-14

Scale 1 to 7 (best)

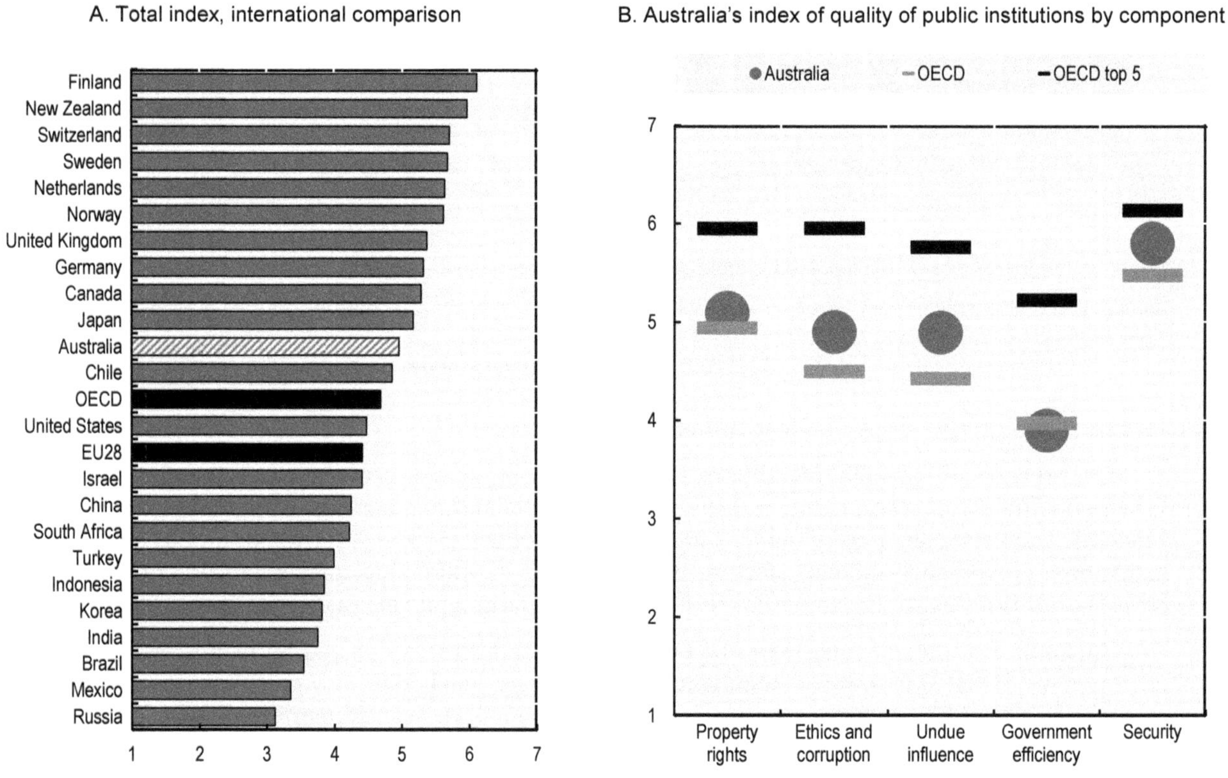

Indexes for EU-28 and OECD represent simple averages of member-country indexes.

OECD top 5 refers to the average of the scores for the top five performers among OECD countries - Finland, New Zealand, Switzerland, Sweden and the Netherlands.

Source: World Economic Forum (2013), *The Global Competitiveness Report 2013-2014*, http://reports.weforum.org/the-global-competitiveness-report-2013-2014/#.

StatLink http://dx.doi.org/10.1787/888933249800

References

Banks, G. (2005), *Structural reform Australian-style: lessons for others?,* Productivity Commission, Presentation to the IMF, World Bank and OECD, May. http://www.oecd.org/australia/39218531.pdf.

EIU (2014), *Country Report: Australia*, September 2014, Economic Intelligence Unit, London.

Gray, E.M., Oss-Emer, M. and Sheng, Y. (2014), *Australian Agricultural Productivity Growth: Past Reforms and Future Opportunities*, Australian Bureau of Agricultural and Resource Economics and Sciences Research Report 14.2, Canberra, February, http://www.daff.gov.au/abares/publications.

OECD (2014), Australia" in OECD Economic Outlook 96 – November 2014. http://dotstat.oecd.org/. .

OECD (2013), *Government at a Glance 2013: Country fact Sheet Australia,* OECD Publishing, Paris. http://www.oecd.org/gov/government-at-a-glance-information-by-country.htm. . DOI: http://dx.doi.org/10.1787/gov_glance-2013-en.

OECD (2012a), *OECD Economic Surveys: Australia 2012*, OECD Publishing, Paris. DOI: http://dx.doi.org/10.1787/eco_surveys-aus-2012-en.

OECD (2012b), *Value for Money in Government: Australia 2012*, Value for Money in Government, OECD Publishing, Paris. DOI: http://dx.doi.org/10.1787/9789264178809-en.

OECD (2011), *Government at a Glance 2011: Country fact Sheet Australia,* OECD Publishing, Paris. http://www.oecd.org/unitedstates/governmentataglance2011informationbycountry.htm. DOI: http://dx.doi.org/10.1787/gov_glance-2011-en

OECD (2010), *The OECD Innovation Strategy: Getting a Head Start on Tomorrow*, OECD Publishing, Paris. DOI: http://dx.doi.org/10.1787/9789264083479-en.

Productivity Commission (2011), *Identifying and Evaluating Regulation Reforms*, Canberra. Productivity Commission Research Report, December. http://www.pc.gov.au/inquiries/completed/regulation-reforms/report/regulation-reforms.pdf.

WDI (2013), *World Development Indicators Database*. World Bank, http://data.worldbank.org/data-catalog/world-development-indicators.

WGI (2013), *Worldwide Governance Indicators database*. World Bank Group, http://info.worldbank.org/governance/wgi/index.asps.

World Economic Forum (2013), *The Global Competitiveness Report 2013-2014*: Full data Edition, Geneva 2013. http://reports.weforum.org/the-global-competitiveness-report-2013-2014/#=.

Annex 3.A1

Key economic reforms in Australia in the 1980s and 1990

It became apparent as of the 1960s that many of policies adopted by the Australian government to pursue social and economic objectives were imposing costs on the economy and constraining income growth.[1] A series of government commissioned reports emphasised the costs of protecting manufacturing industries from import competition, centralised wage determination and government ownership of economic infrastructure. These policies were intended to promote population growth, develop local manufacturing industries and redistribute the gains from natural resources. They also served to make many parts of the economy inefficient, inward-looking and inflexible. A series of reforms over the 1980s and 1990s introduced changes in monetary and fiscal policies, capital markets, trade barriers, industry assistance, taxation, corporatisation and privatisation of government business enterprises, business regulation, labour markets and industrial relations, competition policy, new regulatory arrangements for natural monopoly utilities, and innovation and training:

- *Capital markets* — The Australian dollar was floated in March 1983. Foreign exchange controls and capital rationing (through interest rate controls) were removed progressively from the early 1980s and foreign-owned banks were allowed to compete, initially for corporate customers and, in the 1990s, to act as deposit-taking institutions.
- *Trade and foreign direct investment* — Reductions in tariff assistance began in 1973 with an across-the-board cut of 25%. Further phased tariff reductions began in 1988 and 1991. Other reforms such as the abolition of quantitative import controls — mainly in the automotive, white goods and textile, clothing and footwear industries — gathered pace from the mid-1980s. The effective rate of assistance to manufacturing fell from around 35% in the early 1970s to 5% by 2000. In the 1980s, reforms opening the economy also liberalised foreign investment, increasing competition through the entry of foreign producers into domestic markets.
- *Taxation reform* — Capital gains tax and the dividend imputation system were introduced in 1985 and 1987, respectively. The company tax rate has been lowered progressively from the late 1980s. A broad-based consumption tax (GST) was implemented in 2000, replacing the narrow wholesale sales tax system and a range of state-based duties. At the same time, income tax rates were lowered.
- *Fiscal policy* — From the mid-1980s, fiscal policy targeted higher national savings (and a lower current account deficit) and, from the mid-1990s, concentrated on reducing government debt, primarily financed through asset sales (privatisation). Inflation targeting was introduced in 1993.
- *Labour market policies* — Award restructuring and simplification, and the shift from centralised wage fixing to enterprise bargaining (decentralisation of wage bargaining mechanisms) began in the late 1980s. Reforms accelerated in the mid-1990s with the introduction of the *Workplace Relations Act 1996*, further award simplification (through limiting prescribed employment conditions in enterprise bargaining agreements) and the introduction of individual employment contracts (Australian Workplace Agreements).
- *National Competition Policy* — In the late 1980s, governments began reforms in the key infrastructure sectors of electricity, gas, road transport and water. In 1995, these reform processes were consolidated and broadened in a co-ordinated National Competition Policy (NCP). NCP extended competition into areas of the economy that had been dominated by government monopolies, typically in the provision of infrastructure or where competition was restricted by legislation, including by statutory agricultural marketing arrangements (see Annex 4.A1).

1. This annex draws on Gray et al. (2014), based on Banks (2005) and Productivity Commission (2011).

Chapter 4

Investment in the Australian food and agriculture system

This chapter presents an overview of Australian regulations governing entrepreneurship, access to natural resources and products and processes and discusses the extent to which they affect the adoption of innovative practices and the introduction of new products in the country. It also discusses Australian policies related to trade, investment, finance and taxation and their impact on the capacity of farms and agri-food firms to invest and take advantage of market opportunities.

The statistical data for Israel are supplied by and under the responsibility of the relevant Israeli authorities. The use of such data by the OECD is without prejudice to the status of the Golan Heights, East Jerusalem and Israeli settlements in the West Bank under the terms of international law.

Regulatory environment

Regulations affect innovation through many links and may increase or impede knowledge advancement and technology transfer. Regulations influence the size and behaviour of firms, the ease of entry into and exit from markets, they set standards and impose administrative burden on doing business. All of these factors determine the costs and risks for businesses to innovate.

A broadly recognised link between regulations and innovation is that they determine conditions for competition, and thus incentives for companies to increase productivity. Empirical evidence shows that competition-restraining regulations slow the rate of catch-up with the technological frontier, where labour productivity is highest (OECD 2007, based on Conway et al., 2006). There is evidence that good product market regulation is associated with increased inflows of foreign direct investment and thus technology spill-overs (Nicoletti et al., 2003).

Australia stands out by the importance attached to the quality of the regulatory system as a factor of economic development. A landmark report on National Competition Policy in 1993 recommended substantial reforms to the competition policy framework based on the principle that "there should be no regulatory restrictions on competition unless it was in the public interest..." This resulted in a National Competition Policy Programme (NCP) which included a broad-ranging review and subsequent reform of all legislation restricting competition in all Australian jurisdictions (Annex 4.A1). Through this process, new institutions (e.g. Australian Productivity Commission) and mechanisms were established to support continuous review, impact assessment of regulations, and development of efficient regulations.

A new impetus to regulatory reform was given after the Business Council of Australia (2014) drew attention to an increasing number of new laws across jurisdictions and the re-emergence of inefficient and inconsistent regulations. These issues were taken up by the National Reform Agenda (NRA) launched in 2006. The three streams of this programme are: human capital, competition, and regulatory reform. The latter stream is focused on promoting best-practice regulation and reviewing legislation to reduce burdens on business. All Australian state and territory governments have introduced, or upgraded, regulatory impact systems to improve the scrutiny of new regulatory proposals likely to impose a significant burden on businesses (OECD, 2010b; Gray et al., 2014). In the most recent move, the new Australian government released an independent Competition Policy Review in March 2015.

The OECD Product Market Regulation (PMR) indicators measure the degree to which countries' regulatory framework promotes or inhibits competition in the product markets. They cover key regulations in the areas of state control, barriers to entrepreneurship, and barriers to trade and investment and quantify them as a whole and according to specific dimensions. The integrated PMR indicator for Australia shows that the country's position on regulation in the areas of state control and barriers to business start-ups and foreign trade is overall favourable to competition (Figure 4.1). Nevertheless, regulation in Australia is stricter overall than in some OECD countries, for example, the Netherlands, the United States, or the United Kingdom. Most restraints are related to state control and reflect various forms of state involvement in sectors such as electricity, telecommunications, several aspects of transport infrastructure, water management and water pricing.

Figure 4.1. OECD's Integrated Product Market Regulation (PMR) indicator

Scale from 0 (least) to 6 (most) restrictive

A. Integrated PMR index, international comparison, 2008 and 2013

B. Australia's Integrated PMR index by principal components, 2013

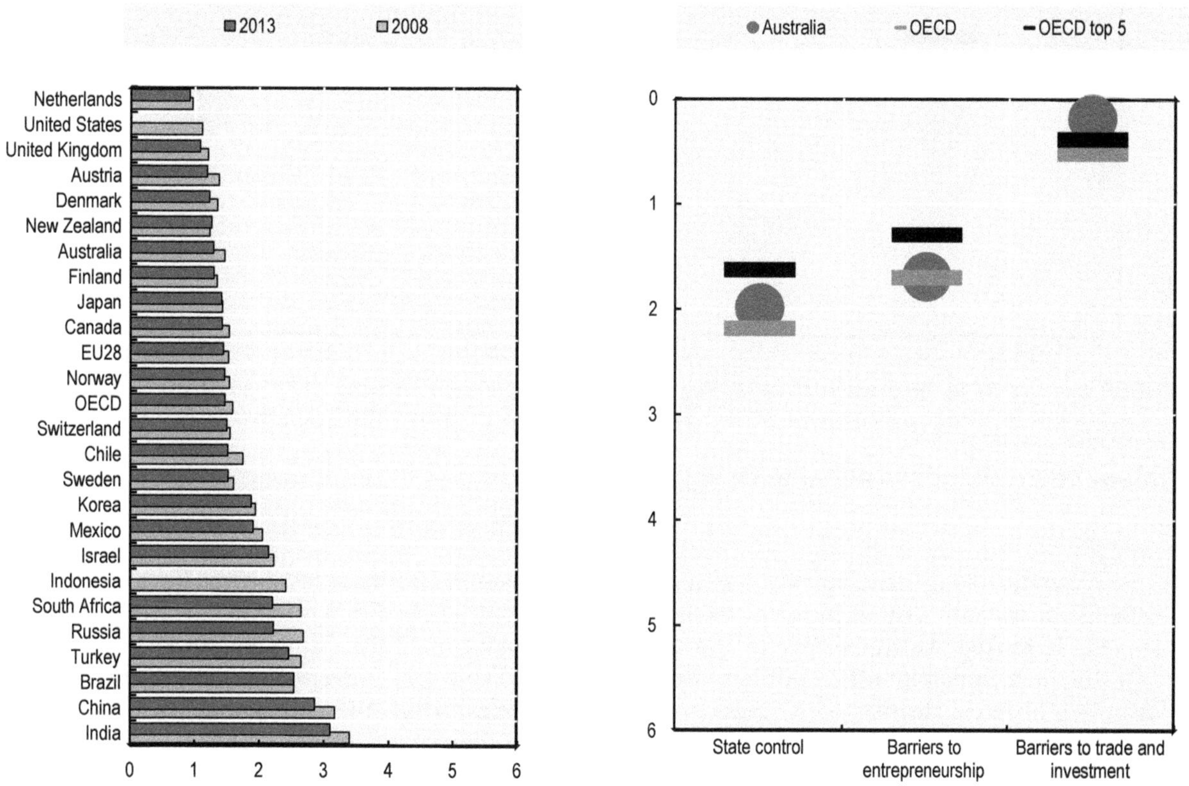

OECD top 5 refers to the average of the scores for the top five performers among OECD countries - Netherlands, United Kingdom, United States, Austria and Denmark. Data for the United States refer to 2008.

Source: OECD Product Market Regulation Database, 2014. www.oecd.org/economy/pmr.

StatLink http://dx.doi.org/10.1787/888933249813

Entrepreneurship regulations

Low barriers to enter markets favour innovation as new companies tend to exploit technological or commercial opportunities which have been neglected by more established companies. While entry and growth of new firms is important, so is their ability to exit and thereby facilitate structural adjustment and the expansion of more efficient companies (OECD, 2010a). Regulatory barriers to entrepreneurship cover aspects such as ease of entry into business and exit from it, the complexity of regulations involving time and costs that businesses spend to comply with these regulations, as well as regulatory protection of incumbents. OECD's PMR indicator estimates Australia's regulations related to all these dimensions to be on aggregate close to the OECD average score (Figure 4.1.B). Concerning individual areas of entrepreneurship regulations, businesses in Australia face the least restrictive regulations on starting business across all 52 OECD and non-OECD countries covered by the PMR analysis. However, Australia's regulatory stance related to protection of incumbents as well as the complexity of regulatory procedures is more restraining than on average in the OECD area (Figure 4.2). Within the component on protection of incumbents, this is driven by certain antitrust exemptions and barriers in network sectors, while Australia's higher than the OECD average score on the complexity of regulatory procedures largely reflects its relatively strict system of licences and permits.

Figure 4.2. Barriers to entrepreneurship indicator for Australia by regulatory area, 2013

Scale from 0 (least) to 6 (most) restrictive

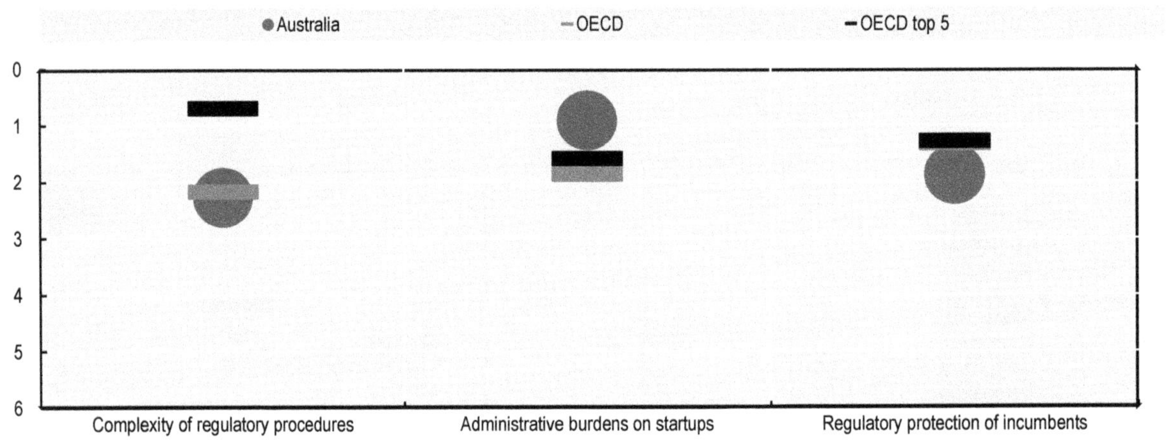

OECD top 5 refers to the average of the scores for the top five performers among OECD countries: Slovak Republic, New Zealand, Netherlands, Italy and the United States on the aggregate barriers to entrepreneurship indicator. Data for the United States refer to 2008.

Source: OECD (2013), *OECD Product Market Regulation Database*. www.oecd.org/economy/pmr.

StatLink http://dx.doi.org/10.1787/10.1787/888933249828

Australia's long-standing efforts to improve and streamline business regulations have enabled it to establish a friendly overall business environment, as reflected by the World Bank's Doing Business survey. It ranked Australia next to the top ten performers in 2014 (tenth in 2013) based on the assessment of important aspects of business functioning (Table 4.1). Nevertheless, this average ranking disguises different degrees of progress across various areas. Thus, the perceptions of local business place Australia among the world's best performers on the conditions for starting a new business (consistent also with the OECD's PMR indicator), on the ease of obtaining credit and construction permits, and contract enforcement. At the same time, businesses see a need for better protection of investors' interests, a simpler tax regime, and easier cross-border trading.

Rural businesses in Australia are governed by around 90 Commonwealth Acts, as well as state and territory legislation, with complex divisions of regulatory power between national and sub-national levels of governance (Annex 4.A2). In 2007, in the context of the National Reform Agenda, the Productivity Commission undertook a comprehensive review of the regulatory burdens on primary sector businesses (Productivity Commission, 2007). It concluded that farmers face a heavy burden of regulations and identified a set of concerns regarding the regulations applied to rural industries. Recently, ABARES re-examined and updated the 2007 Productivity Commission's review with the aim of identifying areas of unnecessarily burdensome regulations which, if improved, could raise productivity of rural industries. Twenty policy areas covering 32 regulatory issues related to agriculture and forestry were considered (Gibbs et al., 2013).[1] The following sections highlight several regulatory issues raised in this review that are relevant from the perspective of agricultural sustainability and innovation and drawing extensively on ABARES work mentioned above. It should be highlighted that beyond the issues presented below, a number of other issues were raised during the most recent public discussions on future agricultural policy. Among these are consistency of animal welfare standards, land planning and environmental requirements, Country of Origin Labelling and others.

Table 4.1. Australia's ranking in Word Bank's Ease of Doing Business, 2014

Topic	Australia's rank out of 189 economies	Australia's distance to frontier (100 = best performer economy)[1]
Overall score	11	80.67
Starting a business	4	95.91
Dealing with construction permits	10	85.41
Getting electricity	34	78.58
Registering property	40	77.64
Getting credit	3	93.75
Protecting investors	68	56.67
Paying taxes	44	78.69
Trading across borders	46	78.28
Enforcing contracts	14	76.28
Resolving insolvency	18	86.14

1. The "distance to frontier" measure shows the distance of each economy to the "frontier," which represents the highest performance observed on each of the topics across all economies included in Doing Business. An economy's distance to frontier is indicated on a scale from 0 to 100, where 0 represents the lowest performance and 100 the frontier.

Source: World Bank (2013), *Doing Business 2014*, World Bank and the International Finance Corporation, World Bank Publications, Washington, DC. http://www.doingbusiness.org/reports/global-reports/doing-business-2014.

Regulations on natural resources

State and territory governments have the main authority in water, land, and air management in Australia. At the same time, increased awareness of environmental, economic and social considerations affecting the society has led to the emergence over time of important nation-wide legal frameworks. The Australian government also assumed a greater role in co-ordination of environmental regulation across jurisdictions and promoting inter-governmental co-operation on environmental matters.

Environment Protection and Biodiversity Conservation Act 1999

This Act is the primary national legal framework for environmental protection and management. Among other things, it applies to nationally and internationally important flora, fauna, ecological communities and heritage sites. It empowers the Australian government to undertake environmental assessments of activities likely to have a significant impact on these assets. Approval is required for proposed actions (including on private land) that are likely to have a "significant impact". If this is the case, a detailed environmental assessment of the action must be conducted to determine whether approval will be granted and under what conditions (if any).

In its 2007 review, the Productivity Commission recorded various concerns on the burden related to the Environment Protection and Biodiversity Conservation Act, but specifically singled out the lack of clarity about what constitutes a "significant impact." Such ambiguity leads to difficulties in interpreting what it constitutes in practice and may result in a farmer foregoing profitable business incurring costs by undertaking actions that may be determined as having a "significant impact" by the regulator.[2] Since the 2007 review, there have been efforts to increase farmer understanding about actions that are likely to require referral to assessment and to provide guidance on the interpretation of "significant impact". ABARES assessment concludes there still appears to be scope for reducing uncertainty among industry, as well as the cost of complying with regulatory requirements. The Australian Department of the Environment is currently working to improve the cost of compliance with its assessment and approval process. In particular, the average assessment time for projects has been halved over the last two years. There is also work to implement the One-Stop Shop policy to reduce duplication between the Australian Government and states and territories in the environmental assessment and approval processes. This is to be achieved, along with maintaining the environmental standards under the *Environment Protection and Biodiversity Conservation Act 1999.*

National Environment Protection Measures and National Pollutant Inventory

The National Environment Protection Measures is a set of nationally consistent environmental standards, goals, or protocols related to six areas: air, water, noise, site contamination, hazardous waste and recycling. The National Pollutant Inventory (NPI) is part of this system that, among other activities, covers agriculture. The NPI collects and disseminates information on the emission and transfer of pollutants from individual facilities to air, land and water, and the transfer or movement of these substances on or off-site. It contains data on 93 substances that may affect human health or the environment. Within agriculture, the NPI requires emissions and transfer data in particular from poultry farms, piggeries and beef cattle feedlots. The information required for specific intensive livestock facilities includes: stock numbers, fuel use and the amount of water that flowed from waste water storage to surface water. In contrast, farmers are not required to report diffuse emissions or transfers from tree growing, aquaculture, horticulture and extensive livestock grazing. These measurements are difficult to be done accurately at the farm level, and state and territory environment protection agencies provide aggregate estimates of diffuse emissions for the NPI. In 2011/12, 517 agriculture facilities reported to the NPI, with 34 different substances reported to have been emitted and three substances transferred.

Concerns were expressed during 2007 Productivity Commission review about NPI's burdensome reporting requirements, involving much time for calculating emissions and complicated data lodgement processes; there have also been suggestions to shift some of the reporting burden to industry groups. Since 2007, measures have been taken to reduce this burden, including simplifying the process by which emissions and transfers are calculated and reported, for example by updating estimation technique manuals and the development of web-based reporting. According to ABARES, industry groups that previously raised this issue no longer consider it important. However, regular reviews of reporting thresholds can maintain pressure to reduce unnecessary reporting burden; greater involvement of industry groups may also help to reduce burden on individual reporters.

Water management

State and territory governments are primarily responsible for regulating water resources within their boundaries. This largely concerns water allocation between users, water infrastructure management and, since the 1980s, environmental and sustainability issues. The Australian Government does not have explicit constitutional responsibility for water resources, but several sections in the Constitution allow the Australian Government to legislate with respect to water, including the external affairs power. The *Water Act 2007* (Water Act) and the (Murray-Darling) *Basin Plan 2012* provides for a Basin-wide approach to the long-term sustainable management of Basin water resources and establishes an independent Murray-Darling Basin Authority (MDBA) with the functions and powers, including enforcement powers, needed to ensure that Basin water resources are managed in an integrated and sustainable way.

The *Intergovernmental Agreement on a National Water Initiative* (NWI) was signed in 2004 by the Australian, state and territory governments to implement water reform. The NWI covers the fundamental aspects of water management, such as: water access rights; planning; water markets; water pricing; integrated water management; water for environmental outcomes; and implementation of the National Water Quality Management Strategy. A key initiative under this reform framework was the establishment of clearly defined and more secure water access rights (known as entitlements). The NWI required the states and territories to separate water access entitlements from land titles and establish rules for trading water within their boundaries. Individual water access entitlements were defined as "shares of the consumptive pool", i.e. water available in a specified area for consumption, rather than a fixed quantity. Water registers record ownership of water access entitlements and trades. Another NWI feature is the principle of cost recovery of services where possible, including the cost of environmental externalities (OECD, 2008).

Substantial progress has been made across Australian states and territories in implementing the NWI. Although the NWI is to achieve a more nationally cohesive approach of how water is managed, priced and traded, the 2007 Productivity Commission review, and more recent reviews, such as a report by the 2011 National Water Commission (NWC, 2011), drew attention to inconsistencies of water access regulations across jurisdictions. The following assessment by ABARES concluded that there appears to have been little progress since 2007 towards a nationally-consistent regulation of water rights (Gibbs et al., 2013). For example, the legislation governing water entitlements varies in unit of measurement, duration and security between jurisdictions. The lack of consistency increases uncertainty in transfer of water rights, adds to administrative burdens faced by businesses through higher compliance costs and may result in lost water use opportunities at critical decision points. According to ABARES, the (Murray-Darling) Basin Plan may remove some regulatory inconsistency across participating jurisdictions. Thus, water trading rules came into effect as of 1 July 2014 under the Basin Plan to enhance water trading across that large and agriculturally productive area. However, there remains scope to standardise the definition of water access rights and to improve performance reporting. An assessment of progress in implementing the NWI is undertaken by the Australian Government every three years.

Regulations on products and processes

Genetically modified crops

Activities related to genetically modified organisms (GMOs), including developing, breeding, growing and transporting GMOs, fall under the *Commonwealth Gene Technology Act 2000* and corresponding state and territory legislation. These regulations are administered by the Gene Technology Regulator, which makes decisions on the development and use of GMOs based on health, safety and environmental considerations but without considering any social or economic dimensions. Commercialisation and marketing arrangements of GM crops are regulated by the states, which have the discretion to impose moratoria on growing GM crops commercially within their territories. The scope of commercially allowed GM crops also varies across the states, from one specific type to all GM crops.[3]

Earlier access to new technologies could enable innovative farmers to realise the benefits from adoption sooner. Several issues related to GM crop commercialisation were expressed by stakeholders in this respect. The imposition of state moratoria delayed this process and eroded national regulatory consistency. The timeliness of approval for commercial cultivation varied substantially across states and territories and sometimes took several years. The assessment by ABARES concluded that with respect to these features, the commercialisation of GM crops continues to carry unnecessary burden for agricultural businesses. Although addressing these concerns is a prerogative of the states, the Australian Government could play a role in providing co-ordination and information in order to shorten the regulatory path and develop greater consistency across jurisdictions.

Agricultural chemicals and veterinary medicine

The regulation of agricultural chemicals and veterinary medicine supply, up to and including the point of retail sale, is under the responsibility of the Australian Pesticides and Veterinary Medicines Authority (APVMA). The use of products beyond the point of retail sale is controlled by state and territory authorities.

All agricultural and veterinary chemicals supplied in Australia must be registered by the APVMA. One issue raised by the 2007 Productivity Commission's regulatory review (Productivity Commission, 2007) is the timeliness and complexity of national chemicals registration and approval procedures. Concerns were also raised about the APVMA's capacity to meet statutory timeframes for registering products. The new *Agricultural and Veterinary Chemicals Legislation Amendment Act* (2013) is aimed at addressing these concerns by, in particular, reducing information requirements, increased transparency of the assessment process, simplifying the application process for certain products, and improving APVMA's business practices. In addition, imports of biological agricultural and veterinary products must be allowed by the Department of Agriculture's Biological Imports Program (BIP), which assesses risks from such products to Australia's biosecurity. The issue raised by the stakeholders in 2007 in this regard was the duplication in approval between the two agencies, leading in turn to long lags in product approval, unnecessary costs to businesses and reduced competitiveness of Australian produce. Since 2007, the duplication in assessment requirements appears to have been reduced through increased co-operation between the agencies involved, simplification of application procedures and information sharing.

Another issue concerns access to chemicals for minor uses. Chemical companies face strict labelling requirements and are generally unlikely to recover the costs of generating the required data and registration of the products in Australia because of the country's small market. This reduces the variety of chemicals available to farmers, affecting in particular the horticultural and developing industries. In addition, restricted access to minor use chemicals may discourage investment in new better targeted chemicals, increase the illegal use of chemicals, and may expose Australian exporters to greater testing requirements in import markets. Various initiatives since 2007 have enabled progress in improving access to minor use chemicals, although there continues to be obstacles. At the level of the Australian government, business costs can be reduced by raising the awareness of chemical manufacturers about market opportunities, promoting better use of existing data and data sharing, and exploring opportunities to bring down the cost of compliance with regulations through industry self-regulation.

State and territory governments regulate the use of agricultural and veterinary chemicals beyond the point of retail sale. The degree of control differs across jurisdictions with respect to licencing and training requirements for chemicals users, notification and record keeping, administration of codes of practice, and monitoring and enforcement of the conditions of use. In the 2007 Productivity Commission Review, concerns were raised about the unnecessary complexity and inconsistency of these regulations across jurisdictions, thus increasing the costs of compliance for farmers. In view of these concerns, the Australian, state and territory ministers for primary industries decided in 2013 to implement a national framework to harmonise the regulation of agricultural and veterinary chemicals and to identify further areas for reform.

Biosecurity and quarantine

Australia, as an island, is free of many pests and diseases present elsewhere in the world, which provides the country significant advantage in access to foreign agro-food markets. Biosecurity is therefore an area of particular importance. The country's biosecurity system manages pest and disease risks across a biosecurity continuum where there is an emphasis on prevention, preparedness and early intervention of pests and diseases with activities undertaken offshore, at the border and onshore. The Australian Government is empowered to manage biosecurity risks at the border and offshore. The management of risks within the country's borders is the responsibility of Australian states and territories. The Australian government, however, has a role in emergency responses to disease outbreaks through such arrangements as the *Emergency Plant Pest Response Deed*, the *Emergency Animal Response Deed*, and the *National Environmental Biosecurity Response Deed*. These arrangements outline emergency response and cost-sharing principles, and complement the activity of state and territory governments in managing biosecurity within the country's borders.

Biosecurity related to the importation and exportation of goods consists of risk assessments and certification, and monitoring and compliance measures. Import risk assessment is a principal element of the biosecurity regime. Through this process, the level of quarantine risk in the importation of plants, animals and other items is determined. If such risk is estimated to exceed the acceptable level after possible measures were taken to reduce it, trade is not permitted. One concern raised by the 2007 Productivity Commission review was the timeliness of import risk analysis, which can take years. Considering this process to be unnecessarily long, the review noted that this creates uncertainty and may impede innovation and reduce incentives to invest in industries potentially exposed to increased competition. Since the Productivity Commission's review, the Australian Government has introduced specified time frames for import risk analysis. However, the length of the process remains a concern. Determining whether it is appropriate to further reduce the time taken to complete import risk analysis would require additional consideration in consultation with the industry.

As concerns the domestic aspects of biosecurity, a lack of national consistency in plant biosecurity regulation has been recognised as an impediment to businesses trading across jurisdictions. A ten-year National Plant Biosecurity Strategy was introduced in 2010 and endorsed by the Australian government, state and territory governments, and industry. It is strongly focussed on achieving greater inter-jurisdictional integration of plant biosecurity system, including adoption of nationally consistent plant biosecurity legislation, regulations and approaches within each state and territory.

Overall, a number of significant reviews have been undertaken in the past two decades outlining opportunities on how Australia's biosecurity system could be improved.[4] Continuous reform of Australia's biosecurity system is being undertaken to ensure the system keeps up with the changing biosecurity environment. A central part of these reform activities has been the focus on managing risks across all points in the biosecurity continuum (offshore, at the border and onshore) and to align regulatory efforts with the level of risk. This approach is enhanced by actions to ensure sustainable funding, improve legislative frameworks, modernise information and communications technology (ICT) systems, and enhance the management of established pests and diseases.

Food regulation

Australia's food regulation system is established by the *Food Regulation Agreement* (2000), which commits the Australian Government and all state and territory governments to a national approach. State, territory and local governments are primarily responsible for implementing and enforcing food standards, while the Australian Government has a role in enforcing food standards at the border. Since 1991, the Australian food regulation system operates in partnership with New Zealand, whereby the two countries have responsibility in developing and maintaining the *Australia New Zealand Food Standard Code*.

One issue of food regulation relates to the considerable freedom that Australian state and territory governments have in implementing a national approach to monitor and enforce food standards. This leads to inconsistencies in food regulation between jurisdictions and increases compliance costs for businesses operating across jurisdictions. There have been efforts at harmonisation, e.g. the introduction of a *Model Food Act* with *Model Food Provisions* to support the development and consistent implementation of standards, but inter-jurisdictional differences remain. The recent assessment by ABARES concludes that unless specific safety outcomes require variations, divergence from the model legislation can create unnecessary burdens. Although the Australian Government has no constitutional power to regulate food, it could facilitate a more nationally consistent regulation, e.g. by revisiting the provisions of the Model Food Act.

Trade and investment policy

Trade openness and foreign direct investment (FDI) are critical parts of the framework conditions for innovation. The expansion of markets worldwide has been a main driver behind technological innovation and productivity gains as larger-sized markets have become available to innovators and consumers (OECD, 2010c). Countries that have shown high innovation performance commonly based their growth strategies on opportunities coming from international markets. Besides increasing external markets, international openness results in more competition domestically and thus increases the pressure on companies operating on the domestic market to innovate in order to hold up to competition.

In addition to creating market size and acting as competition drivers for innovation, trade and FDI operate as immediate channels of technology, know-how and managerial expertise, and have indirect effects on innovation. Inward FDIs not only bring innovations to the businesses directly involved, but may have spill-over effects on other companies in the same industry. This may come through a competition effect when domestic businesses improve their processes and products in response to FDI; through demonstration effect when domestic actors imitate better practices of companies with foreign capital; or through labour market effects when training of local workers prompts a learning process that can, with time, reach out to the rest of the economy. FDI spill-overs may also go through the businesses upstream or downstream the firm with FDI which may face the need to meet new standards to adapt to the requirements of the foreign firm (Havarnek and Irsova, 2012).

Australia's overall economy exposure to trade is relatively modest – a sum of exports and imports of all goods and services is around 40% of GDP, according to the most recent available estimate (OECD, 2014). This is one of the lowest levels across the OECD area, reflecting to a certain extent Australia's remoteness from some key world markets. This share, nevertheless, has tended to increase since the start of the commodity boom which boosted the exports of minerals, such as coal and iron ore, to Asian markets.[5] Trade flows in both directions are roughly comparable in value, yielding a narrow trade balance which periodically switches from positive to negative. Over two-thirds of Australian merchandise exports in 2012 were directed to East Asia (China, Japan, India and Korea); these markets in turn accounted for over one-third of the country's imports. On the import side, Australia is substantially integrated with the European Union and the United States, which account for nearly the same aggregate share as the Asian suppliers (WTO, 2014).

The integration of the economy with international markets, apart from geographical characteristics and the size of markets, is facilitated or impeded by trade policies. Protection of domestic markets through tariff and non-tariff barriers weakens competitive pressure on local producers and thus affects their incentives to innovate and their capacity to do so if protection relates to capital and intermediate goods.

The WTO characterises Australia as one of the most open economies in the world (WTO, 2011). Trade openness has been an important pillar of the policy to enhance competitiveness and productivity of the Australian economy. Since the 1980s, the government has been implementing unilateral tariff reductions in merchandise trade and further steps are envisaged by 2015. Barriers to trade in services were also significantly reduced. Tariff protection remains one of the main trade policy instruments, with an average applied MFN tariff in 2012 at 2.7% and 97% of total imports facing tariff rates within zero to 5% range (WTO, 2014).

Australia's trade relations combine multilateral, regional and bilateral approaches. Seven comprehensive Free Trade Agreements (FTAs) are in force, both regional and bilateral, accounting for 28% of Australia's total trade. These include FTAs with New Zealand (ANZCERTA 1983), Singapore (SAFTA 2003), Thailand (TAFTA 2005), the United States (AUSFTA 2005), Chile (Australia-Chile FTA 2009), the ASEAN-Australia-New Zealand Free Trade Area (AANZFTA 2010) and Malaysia (2013). In 2014, the Korea-Australia Free Trade Agreement (KAFTA) was officially approved and the official approval of the Japan-Australia Economic Partnership Agreement (EPA) is nearing completion. New bilateral negotiations are underway with China, India and Indonesia and Australia is engaged in several plurilateral FTA negotiations, including the Trans-Pacific Partnership Agreement (TPP).

The OECD PMR index of regulatory restrictions to trade and investment translates into single numbers countries' provisions on tariff protection, differential treatment of foreign suppliers, barriers to FDI and barriers to trade facilitation. Australia has one of the least restrictive frameworks covering these key policy areas. It also has been making important moves in trade and FDI liberalisation, as indicated by more than halving the index from 0.53 in 2008 to 0.19 in 2013, the greatest reduction across the monitored countries during this period (Figure 4.3.A). At present, Australia's overall score on regulatory restrictions to trade and investment is determined by the restrictions existing in the FDI area (Figure 4.3.B).

Figure 4.3. Index of regulatory restrictions to trade and investment, 2008 and 2013

Scale from 0 (least) to 6 (most) restrictive

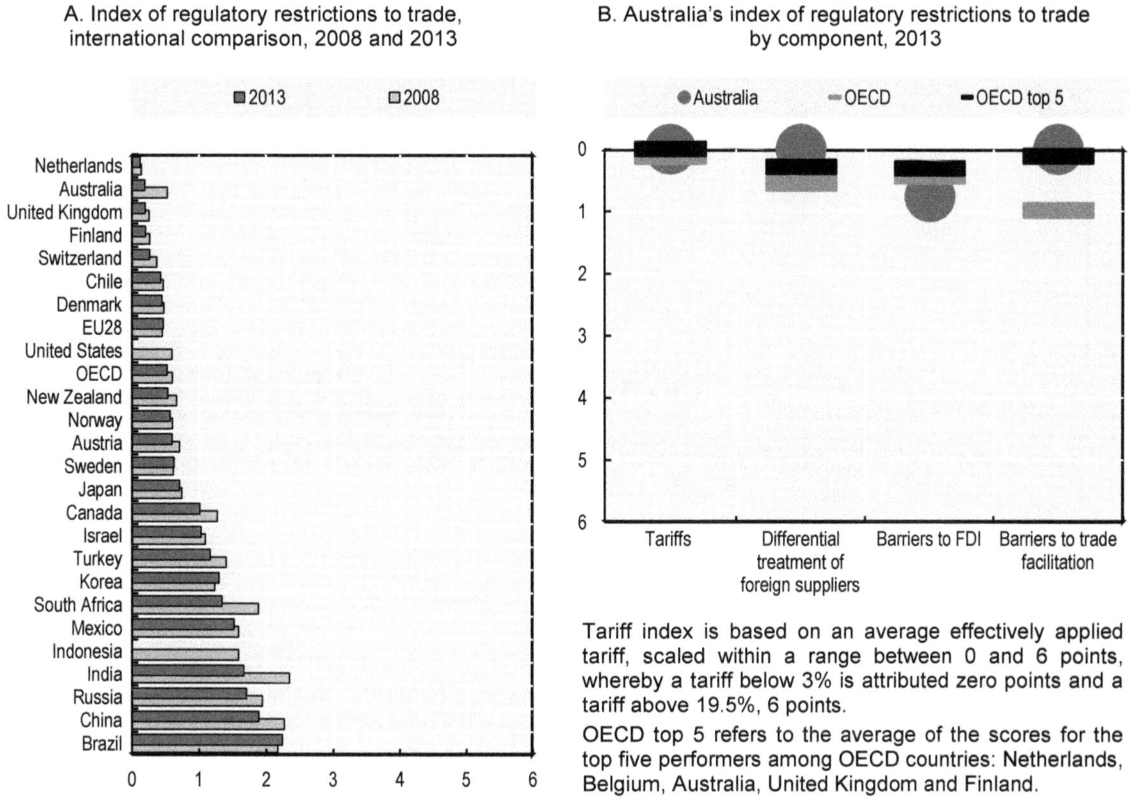

Tariff index is based on an average effectively applied tariff, scaled within a range between 0 and 6 points, whereby a tariff below 3% is attributed zero points and a tariff above 19.5%, 6 points.

OECD top 5 refers to the average of the scores for the top five performers among OECD countries: Netherlands, Belgium, Australia, United Kingdom and Finland.

Source: OECD Product Market Regulation Database, 2014. www.oecd.org/economy/pmr.

StatLink http://dx.doi.org/10.1787/888933249828

Australia has long relied on FDI to increase investment in the domestic economy. Furthermore, FDI has historically been a way of bringing technology into Australia, with the major part of FDI inflows originating in countries with a high level of technological development, such as the United States, Canada, and the United Kingdom (Sanyal, 2014).

FDI liberalisation was an important facet of the major reforms in the 1980s and 1990s, and further simplification and relaxation of the regime occurred in 2008-09. FDI was also covered by the regulatory reforms in Australia that were aimed at raising the efficiency and coherence of regulatory procedures. Today, Australia's foreign investment policy is generally open and transparent, but not without restrictions. The OECD FDI Regulatory Restrictiveness Index evaluates the country's regulations to be more restrictive than on average across the OECD area (Figure 4.4). This index covers 22 economic sectors in each country and accounts for several key aspects of the FDI regime, such as the presence of limitations on foreign equity, screening and approval mechanisms, restrictions on employment of foreigners, and operational restrictions on foreign companies. Some of these restrictions exist in Australia and have cross-sectoral coverage, thus applying to the agriculture and agro-food sectors.

Figure 4.4. OECD FDI Regulatory Restrictiveness index, 2013

Scale from 0 (least) to 1 (most) restrictive

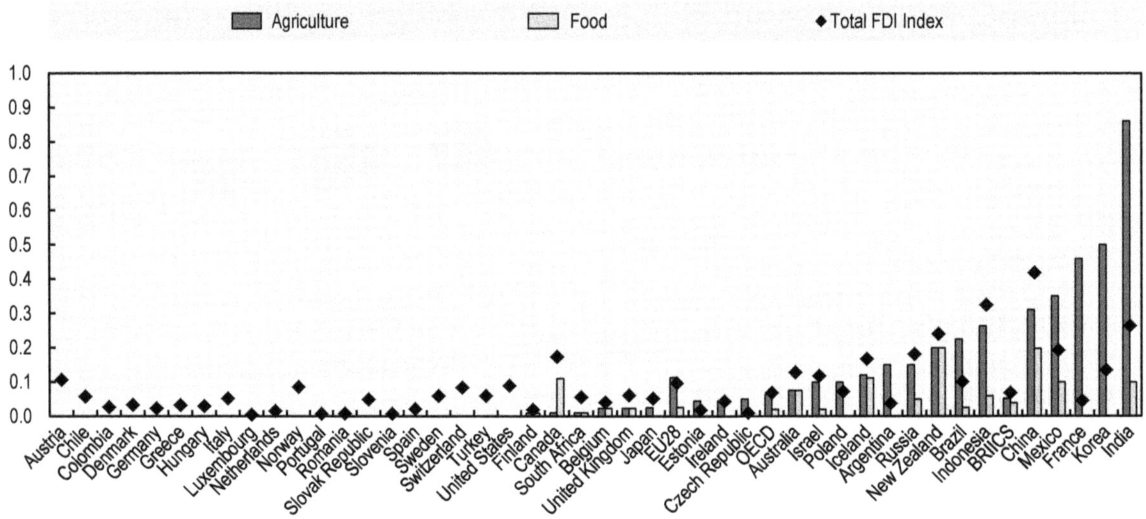

Source: OECD FDI Regulatory Restrictiveness Index. http://www.oecd.org/investment/fdiindex.htm.

StatLink http://dx.doi.org/10.1787/888933249842

Some foreign equity restrictions exist in sectors considered as sensitive.[6] None of them apply to the agro-food and agriculture sectors, or agricultural land acquisitions. However, at the state level, an authorisation is required in Western Australia for the transfer of pastoral leases to ensure a majority of Australian ownership. The broadest provision that concerns all sectors is the requirement to receive approval for all inward foreign investment with a value beyond a certain threshold. Depending on the value of the investment, such approval is granted with or without detailed screening.[7] In the latter case a judgement by the Treasurer is necessary that such investment is not contrary to the national interest. The Treasurer can block foreign investment proposals, or can impose conditions or undertakings on an investment to address national interest concerns (OECD, 2010b; OECD, 2013).

Agriculture constitutes a small share of FDI stock in Australia, accounting for only 0.12% of the total in 2011. At the same time, this sector has been one of the fastest growing FDI destinations. Approved FDI inflows in agriculture increased from AUD 8 million in 2005/06 to AUD 2.9 billion in 2012/13 and their share in total approved FDI inflows[8] increased from 0% to 2.1% (Sanyal, 2014; FIRB, 2014). FDI inflows in agriculture led to a marked expansion of agricultural establishments with foreign capital, which now concentrate 11% of total agricultural land (mostly in mixed farming) and 9% of water rights (Table 4.2). Foreign owners of agricultural land include agribusiness companies that integrate upstream the supply chain and various entities not involved in agriculture, such as pension funds and mining companies seeking to diversify their investment portfolios.

Table 4.2. Agricultural assets in partial or full foreign ownership in Australia

	March 1984	June 2013
Agricultural establishments with joint and foreign control, % of total number	0.3	0.6
Agricultural land partly and full foreign-owned, % of total area	5.9	12.4
By sector		
Nurseries and floriculture	0.2	1.5
Vegetables, mushrooms	0.1	1.4
Fruit and tree nuts	0.8	2.3
Grape growing	4.4	13.2
Mixed beef, sheep, and grain farming[1]	6.1	13
Beef feedlot	..[2]	..[2]
Other crops	4.8	9.3
Dairy cattle	0.3	2.4
Poultry	0.9	1.5
Deer, other livestock	0.1	2.2
Other	0.3	6.0
Water entitlements partly or fully foreign-owned, % of total	..	13.7

.. not available.

1. Mixed beef, sheep and grain farming accounts for 94% of total agricultural land.
2. Data not available for publication but included in totals where applicable.

Source: Moir (2011) based on Australian Bureau of Statistics data for 1984 data; Australian Bureau of Statistics, Agricultural Land and Water Ownership Survey, June 2013.

Table 4.3. Share of largest firms with partial or full foreign ownership

Industry	Activity	Share of largest firms	
Grain marketing	Grain purchases from farmers	Cargill (United States)	18
		Viterra (Canada)	12
		Emerald (Australia/Japan)	8
		Elders/Toepfer (Australia/Germany)	5
		Glencore (Switzerland)	4
		Total of five companies	**47**
Meat processing	Red meat throughput	JBS (Brazil°	25
		Cargill (United States), half of Teys/Cargill	8
		Nippon (Japan)	6
		Harmony (Cayman Islands)	3
		Total of four companies	**42**
Milk processing	Raw milk processing	National Foods (Japan)	25
		Fonterra (New Zealand)	21
		Parmalat (France)	7
		Total of three companies	**53**
Sugar milling	Raw sugar production	Sucrogen (Singapore)	43
		Bundaberg (Belgium)	11
		Tully (China)	5
		Total of three companies	**59**

Source: Moir (2011) based on various sources dated between 2010 and 2011.

Australia has seen significant FDI in agro-food industries. The principal investors are companies involved in the same or similar business in other countries. The data on FDI stock in agro-food industries is not available, but approved FDI inflows in these sectors in recent years were roughly comparable with those in agriculture. In 2009/10, they were estimated at AUD 3.4 billion and accounted for 2% of all approved FDI inflows. A small number of companies with partial or full foreign ownership currently account for nearly, or over one-half of business activities in key agro-food segments in Australia, particularly in milk and sugar processing (Table 4.3).

Finance policy

Innovation typically requires borrowing or other types of external funding, particularly by start-up businesses. A well-functioning domestic financial system with sufficient provision of varied services to borrowers of different profiles facilitates the innovation process. As innovation usually requires long-term investment, a strong long-term finance segment is of critical importance. An adequate domestic financial system is important from the perspective of innovating SMEs as they are likely to depend more on internal sources of finance compared to large businesses capable of drawing on international funding. This is particularly relevant in the rural context where SMEs dominate.

Australia has one of the fastest growing and most sophisticated financial markets in the Asian region. A large local equity market operates in the country, with the turnover of debt securities and stocks reaching AUD 4.1 trillion in 2012/13, over-the-counter and organised trade combined (AFMA, 2013). However, the scale of the equity market in Australia is somewhat smaller than in the United States or the United Kingdom, as evidenced by the level of market capitalisation of listed companies and stocks traded relative to GDP (Figure 4.5).

Figure 4.5. Selected indicators of financial markets, 2012

As a percentage of GDP

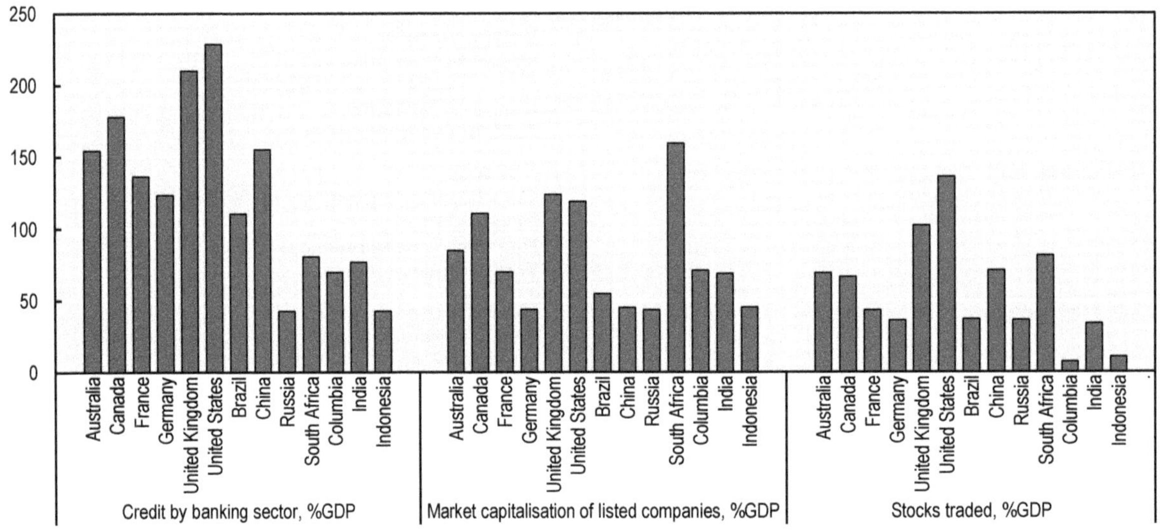

Source: WDI (2013), *World Development Indicators Database*. World Bank, http://data.worldbank.org/data-catalog/world-development-indicators.

StatLink http://dx.doi.org/10.1787/888933249855

Australia has a well-functioning banking sector, with large, sufficiently capitalised and profitable banks. The robustness of the banking sector was demonstrated during and after the 2008-09 financial crisis.[9] For agriculture, borrowings, mostly from banks, represent the predominant source of external funding. Following the deregulation of the financial sector in the late 1980s, access to bank finance by rural borrowers has increased considerably; this was helped by lower nominal interest rates, stronger competition among lenders that narrowed bank margins and risk premiums on business credit. An increased availability of credit for farm businesses is, to a certain extent, evidenced by the substantial rise in rural debt relative to the agricultural output value, in particular since the mid-2000s. The presence of a well-functioning credit market is also evidenced by its ability to support increased land and other capital acquisitions taking place since 2000.[10] According to the Australian Bankers' Association, prior to the global financial crisis small businesses generally had sufficient access to finance and some banks were easing their credit standards, while non-banking lenders offered credit to segments. The global financial crisis strengthened prudency both among lenders and borrowers, leading to a fall in debt levels from their peak in the late 2000s. Recent trends, however, indicate a new up-turn in rural sector lending (Productivity Commission, 2014).

Figure 4.6. Global Competitiveness Index: Index of financial market development, 2013-14

Scale 1 to 7 (best)

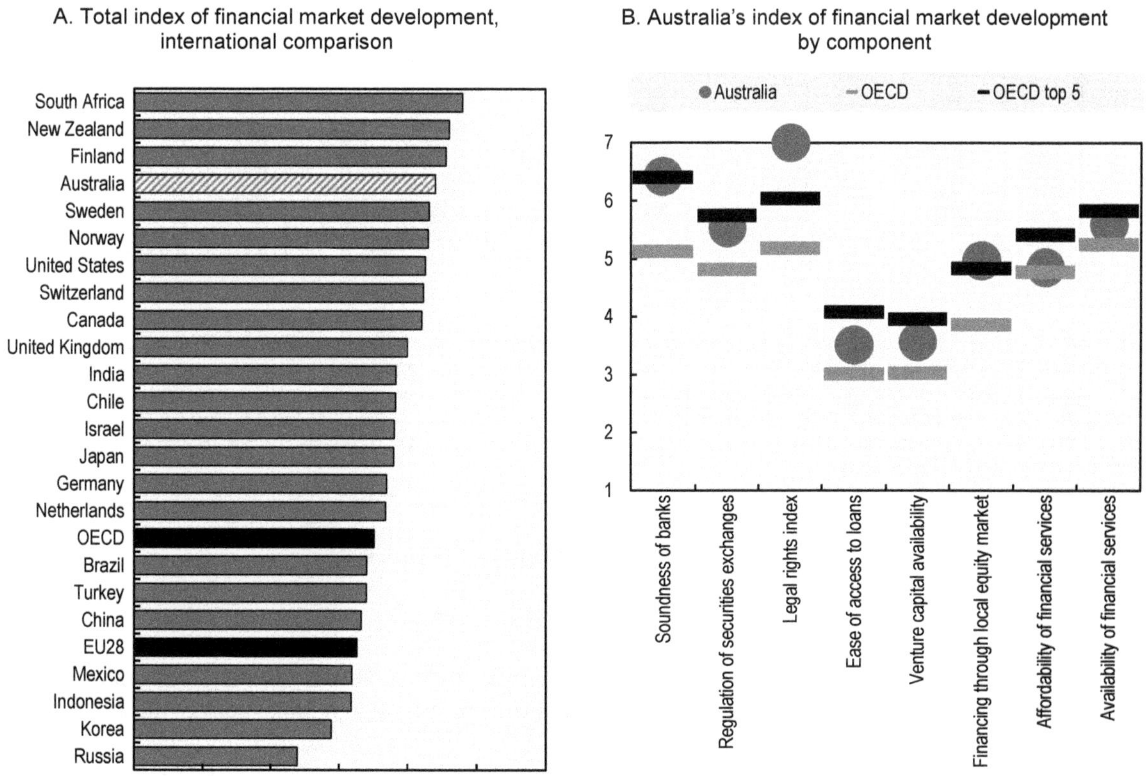

Indexes for EU-28 and OECD represent simple averages of member-country indexes.

OECD top 5 refers to the average of the scores for the top five performers among OECD countries in total index of financial market development - New Zealand, Finland, Australia, Sweden and Norway.

Source: World Economic Forum (2013), *The Global Competitiveness Report 2013-2014*: Full data Edition, Geneva 2013. http://reports.weforum.org/the-global-competitiveness-report-2013-2014/#.

StatLink http://dx.doi.org/10.1787/888933249872

Based on the opinions of Australian businesses, the World Economic Forum scored the country's financial market development at 5.1 points on a scale of 1 to 7 (best) (Figure 4.6A). This is the third highest ranking among OECD countries, after New Zealand and Finland. Australian businesses evaluate all surveyed dimensions of the domestic financial market above the OECD average (Figure 4.6B). The highest appreciation is given to the soundness of banks, opportunities for equity finance, and the system of regulation of securities. Nevertheless, as elsewhere, Australian businesses feel a lack of venture financing and would like to face an easier process for obtaining credit.[11] At the same time, the 2013 World Bank's *Doing Business* evaluation ranks Australia in the top third out of 189 economies for ease of getting credit (Table 4.1). In addition to business perceptions, this evaluation also uses objective information on specific parameters of financial markets.[12]

Figure 4.7. Percentage of surveyed Australian businesses indicating access to finance among the key barriers to innovation, 2012/13

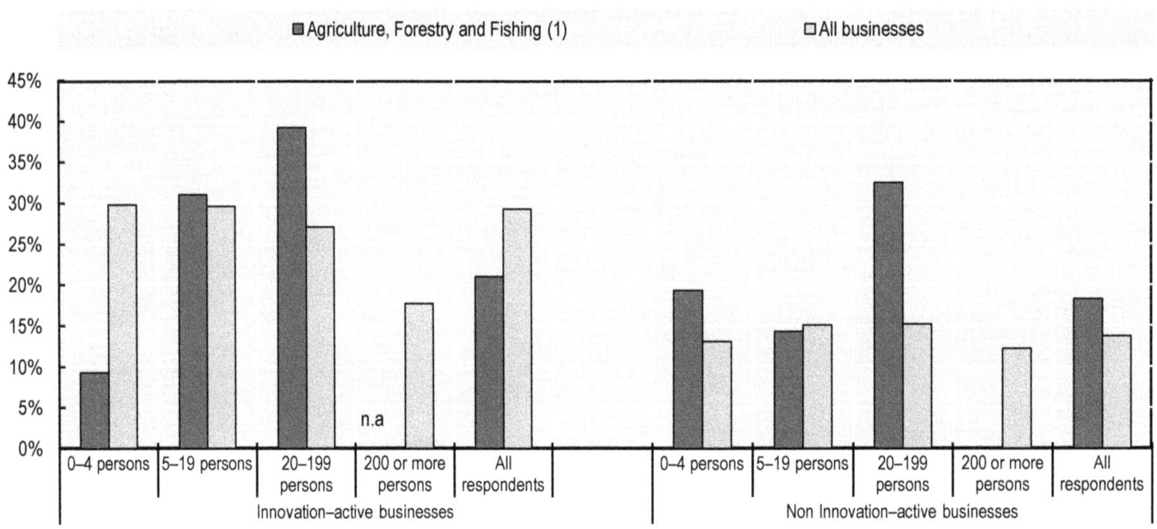

All businesses were asked to identify barriers which significantly hampered the development or introduction of new or significantly improved goods, services, operational or organisational/managerial processes or marketing methods (i.e. innovation related activities). Innovation-active businesses - businesses which had undertaken any innovative activity during the reference period including: introduction of any type of innovation; and/or the development or introduction either still in progress or abandoned; Non innovation-active businesses - Businesses which, in the reference period, did not undertake any innovative activity.

1. Estimates for agriculture, forestry and fishing have a relative standard error of 10% to less than 25% and should be used with caution.

Source: ABS (2014), Innovation in Australia Business 2012/13, cat. No. 81580, August. http://www.abs.gov.au/ausstats/abs@.nsf/mf/8158.0.

StatLink http://dx.doi.org/10.1787/888933249872

The importance of finance as a factor of innovation is underscored by an innovation survey by the Australian Bureau of Statistics (ABS, 2014). Around 19% of respondents across all businesses in agriculture, forestry and fishing, i.e. active and non-active in innovation, named a lack of additional finance among barriers significantly hampering innovation, making it the most frequently reported barrier. This represents roughly a similar proportion if businesses in all economic areas are taken into account, implying that on aggregate agricultural businesses perceive the same degree of difficulty as elsewhere (Figure 4.7). The survey also shows that innovation-active businesses are more sensitive to the availability of additional finance than businesses that are not active in innovation; the former were significantly more frequent to evoke finance constraints as compared to companies that were not active in innovation. However, the results of this survey need to be interpreted with care as they aggregate responses by farming and non-farming businesses (forestry and fishing), the latter most likely dominating the groups with high employment.

The perception of farm businesses of finance as the most frequent constraint to innovation raises the question as to why this is so. Given the highly competitive nature of the rural lending market, and the well-regulated nature of Australian banks, this perception may simply reflect the appropriate pricing of debt finance for risky investments. The results of the survey may also be affected by a particular juncture when following the global financial crisis there was a reassessment of risks associated with business lending and a diminished presence of non-banking lenders that may have affected more strongly projects that involved innovation. It may also indicate the presence of a failure of capital markets to cater to presumably riskier innovative undertakings due to insufficient information and knowledge among lenders to assess the risks involved. This raises another question on whether there is role for government to improve access to finance for innovating farm businesses, for example, by reducing possible information asymmetries amongst rural lenders. Increasing venture finance opportunities for agro-food industries could be an additional venue, in particular to identify the constraints to this type of finance, and a possible government role to ease these constraints.

Tax policy

The principal link between tax policy and innovation is that taxation affects the returns to innovation and thus the decisions of firms and individuals to invest. Taxation also affects the relative price of production factors and therefore priority areas for innovation. Beyond that, taxation often acts as a targeted tool to stimulate innovation, e.g. by providing preferences to private businesses that invest in R&D, offering preferential regimes to young innovative companies, and VAT concessions on innovative products. Furthermore, tax policy can steer innovation towards specific areas, e.g. to address particular societal concerns and towards greener technologies and practices, or environmental R&D. Tax policies can also work on the consumer side of innovation by creating incentives for households to purchase products with particular characteristics, e.g. by providing consumer tax concessions on newly-developed national products or environmentally-friendly goods.

Corporate income tax, tax wedges on wages,[13] and sales tax are the key business taxes. In comparison with the majority of OECD countries, Australian businesses enjoy lower rates of wage and sales taxation, although they face higher income taxation (Figure 4.8). The combined central and sub-central (statutory) corporate income tax rate in Australia (30%) is the seventh highest among OECD countries.[14] An indication of the aggregate burden on profit from an array of taxes imposed on business may be drawn from the World Bank-PwC's *Paying Taxes* survey, which evaluates that a "standardised business" in Australia returns in taxes 47% of the net before tax profit (Figure 4.9). This places Australia somewhere in the middle range among the key OECD and non-OECD economies, but at a substantial distance from the OECD countries with the least burdensome tax regimes for companies, such as Canada, Chile and Korea. In terms of the transactions costs involved in complying with tax requirements, such as the time required to do that and the number of taxes payable, conditions for Australian business are among the most favourable. Companies face a relatively small number of taxes and spend less time to comply than do businesses in key OECD and non-OECD economies (Figure 4.10).

As the World Bank-PwC's business taxation indicators relate to a "standardised business," a number of assumptions about the profile of such a business are employed. Among other assumptions, only companies performing general industrial and commercial activities and which do not participate in foreign trade are covered. Both these criteria narrow the inference that can be made from these estimates with respect to agriculture and agro-processing firms. However, to the extent the latter face a taxation regime similar to World Bank-PwC's "standardised" company, the estimates remain instructive.

Figure 4.8. Key business taxes in Australia, international comparison, calendar year 2012

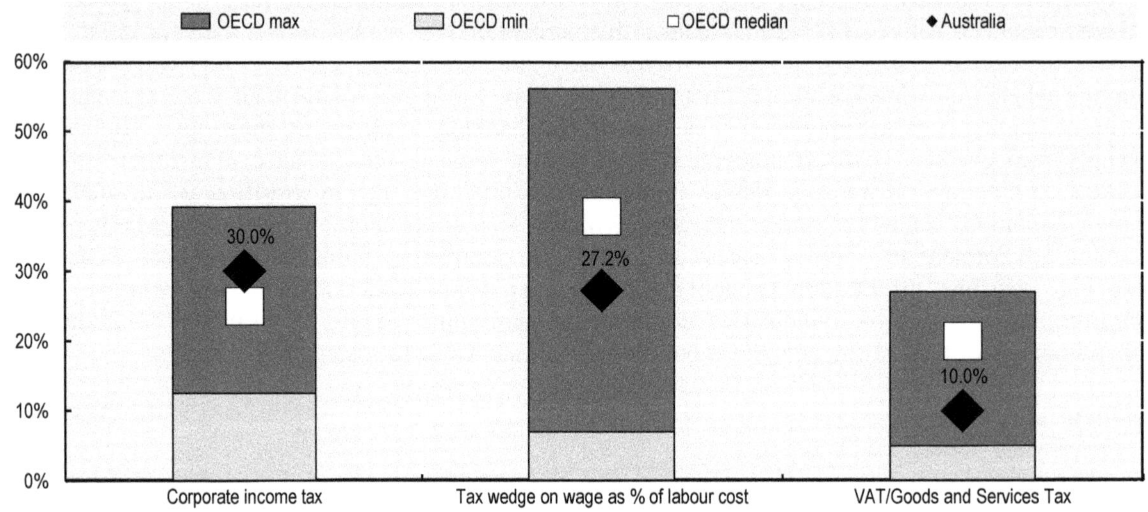

Corporate income tax rate is the basic combined central and sub-central (statutory) corporate income tax rate. Tax wedge on wage includes labour costs paid in income tax and employee and employer social security contributions; OECD aggregate rates for VAT/Goods and services tax do not include data for the United States.

Source: OECD *Tax Database* (2013). http://www.oecd.org/tax/tax-policy/tax-database.htm.

StatLink http://dx.doi.org/10.1787/888933249888

Figure 4.9. Total tax rate on company profits for Australia, international comparison, calendar year 2012

As a percentage of commercial profit

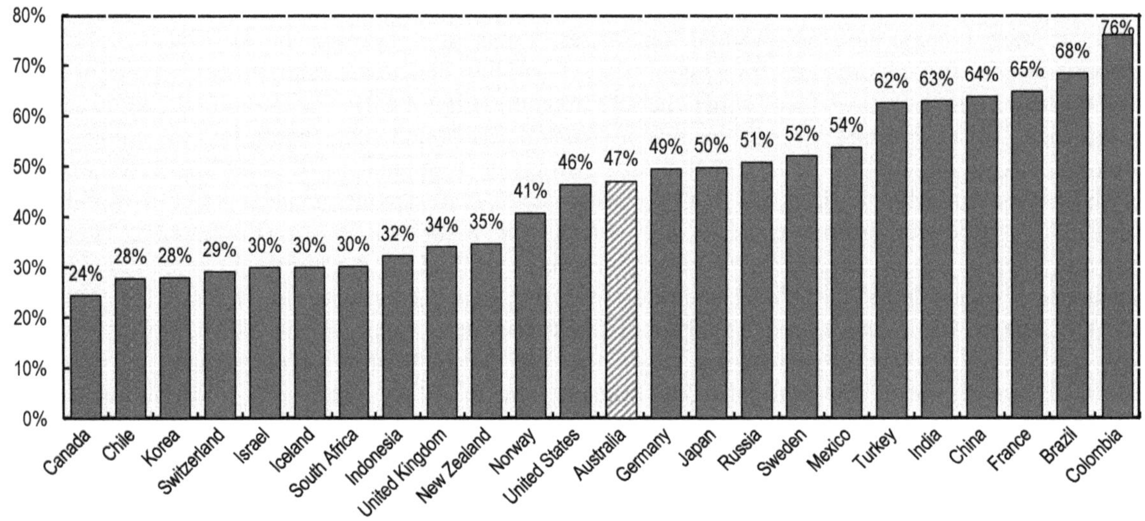

The evaluation uses a concept of a "standardised business" defined on the basis of a set of criteria, including the legal form of business (limited liability), start date of operation (January 2010), geographic location (the largest business city in the country), origin of ownership (100% owned by domestic natural persons), type of activity (general industrial and commercial), size (own capital amount, number of employed, turnover, etc.). The total tax rate is the sum of taxes and contributions payable after accounting for allowable deductions and exceptions related to commercial profit of businesses before all taxes borne. The groups of taxes covered include: profit or corporate income tax; employer's social contributions and labour taxes; property taxes; turnover taxes and other (such as municipal fees, vehicle and fuel taxes).

Source: World Bank Group and PwC (2013), Paying taxes 2014 - The Global Picture, PwC, World Bank and IFC, Washington, DC. http://www.pwc.com/payingtaxes.

StatLink http://dx.doi.org/10.1787/888933249897

Figure 4.10. Number of taxes for a business company and hours required to comply, calendar year 2012[1]

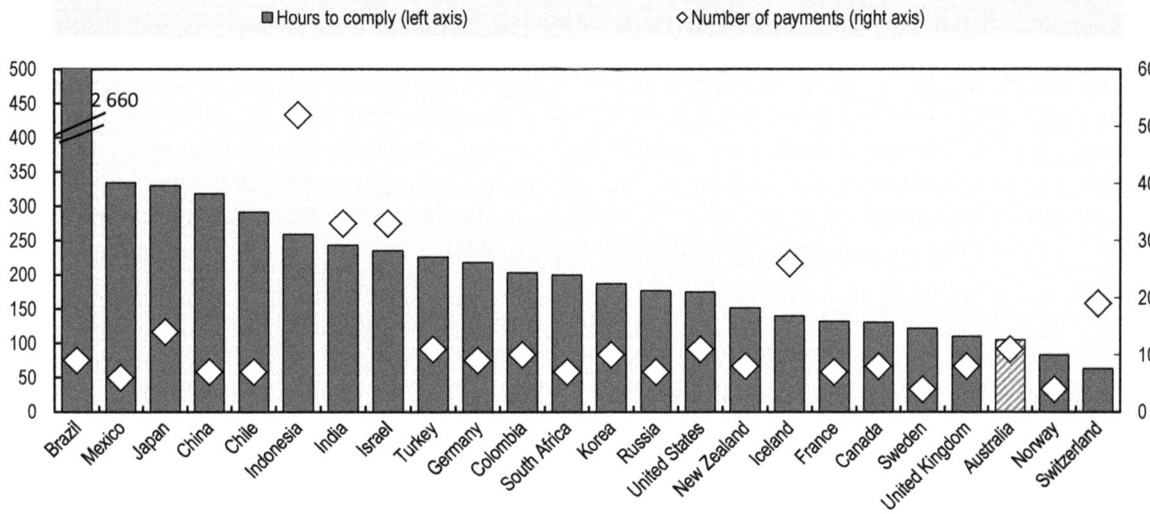

1. See note to Figure 4.9.

Source: World Bank Group and PwC (2013), Paying taxes 2014 - The Global Picture, PwC, World Bank and IFC, Washington D.C. http://www.pwc.com/payingtaxes.

StatLink http://dx.doi.org/10.1787/888933249903

Tax incentives for private R&D is a long-standing policy in Australia and a range of specific schemes related to innovation activity are currently available to businesses (Box 4.1). Australia's tax regime also provides for a range of tax preferences to agricultural producers; these are discussed in the section on agricultural policy (Section 5.1).

Box 4.1 Tax concessions related to private R&D and innovation

The R&D Tax Incentive is the principal mechanism to support private R&D in Australia, offering broadly based and substantial support.1 This entitlement programme is open to corporations of all sizes in all sectors that are conducting eligible R&D. Both Australian and, in certain circumstances, foreign firms with R&D activity in Australia are eligible. An income tax offset (refund) is available to all such companies, which is set as a percentage of their eligible R&D expenditure (Australian Treasury, 2014). Recently, the Australian Government introduced to Parliament a proposal to reduce the rate of this tax offset.

Tax preferences related to innovation are offered to certain investment companies. Establishments that are registered as Pooled Development Funds (PDFs) benefit from a reduced income tax rate, which is set at 15% on investments in small and medium enterprises and 25% on income from other investments (with a standard income rate of 30%). Additional incentives are provided to investors who invest in PDFs; they are not liable for tax either on dividends paid by the PDF or on capital gains on the sale of their shares in PDF. Tax concessions are also provided to certain funds investing in venture capital through the PDFs.

Venture capital operation is supported through individual preferences to its managers. Thus, if the latter are paid performance-based share of partnership profits by investors (i.e. "carried interests"), such payments are treated as managers' capital gain. The taxation of such income may be deferred until the gains are realised and individual managers are eligible for the 50% discount on their carried interest. Certain non-resident investors are exempt from tax on profits and gains on their eligible venture capital investments.

1 The current scheme is a modified successor of an array of previous preferences, including R&D expenditure deductions from taxable income and tax offsets.

Source: Australian Treasury (2010), *Australia's Future tax System: Final Report*, the Treasury, Commonwealth of Australia, January 2010, Canberra. http://taxreview.treasury.gov.au/Content/Content.aspx?doc=html/home.htm.

Since 2010, Australia's tax system has been undergoing significant changes. The previous government began a tax reform that focussed on spreading the benefits of the mining boom throughout the economy. A number of tax initiatives were legislated, including the Minerals Resource Rent Tax (MRRT) on "super-profits" from the mining of iron ore and coal (OECD, 2012). The current government has made the major overhaul of the tax system its principal economic policy commitment. As an immediate move, the MRRT was repealed. This also involved the loss of a number of related tax measures as this tax was intended to provide revenue compensating for the effects of other tax reforms.[15] It was also decided not to proceed with a large number of previously intended tax measures. These and other immediate changes in the tax regime are to be followed by a comprehensive tax reform upon delivery of a White Paper on Tax Reform by the end of 2015.

Notes

1. The ABARES assessment was not intended to cover the full range of regulatory issues and was limited by the regulatory issues which were found to be of particular importance by the Productivity Commission in its 2007 review and, in turn, which the Australian Government accepted or noted.

2. According to Australian Department of the Environment, agricultural activities in practice rarely require assessment and approval under the Act. For example, since the Act commenced in 2000 there have been a total of 54 referrals related to agricultural activities with around 20 requiring assessment and approval.

3. Currently, there are moratoriums on commercial production of GM food crops in Tasmania and South Australia, and, with the exception of canola, in New South Wales and Western Australia. GM canola was approved for commercial release in New South Wales in 2008 and in Western Australia in 2010. The Victorian moratorium on GM canola was allowed to lapse in 2008 and the default provision in Victoria is that all GM food crops approved by the Office of the Gene Technology Regulator may be grown commercially unless an order prohibiting their cultivation is made.

4. These reviews include the Nairn Review (1996) and Beale Review (2008), and to some extent other reviews.

5. Australia's geographical remoteness from world markets leads to an economic handicap relative to most OECD economies. Domestic businesses are limited in exploiting scale economies and this, among other effects, is reflected in trade flows. However, the remoteness handicap is likely to decline as the centre of gravity of world markets shifts eastward to markets that although still distant from Australia, are closer than those of Europe or the United States (OECD, 2012).

6. Air transport cabotage is reserved to Australian-based airlines and 49% equity ceilings exist in three sectors: international aviation, airports offered for sale by the Commonwealth, and domestic shipping (OECD, 2013).

7. The screening process provides for case-by-case assessment of significant proposals. At present, proposals from non-government investors to acquire an interest of 15% or more in a business in a non-sensitive sector valued at more than USD 248 million (or USD 1078 million for United States and New Zealand investors) must be notified to the Foreign Investment Review Board. The factors typically considered in assessing proposed acquisitions in the agricultural sector include the impact of the proposal on the quality and availability of Australia's agricultural resources, Australia's capacity to remain a reliable supplier of agricultural production, biodiversity, and employment and prosperity in Australia's local and regional communities.

8. Approved FDI inflows are an approximate measure corresponding to intended investment; it does not reflect the amount of investment actually realised nor over what time period.

9. Low direct exposure to the euro area, particularly to countries under the strongest financial stress, limited exposure to the mining sector, and the borrowings hedged back to the national currency have contributed to the resilience of Australian banks to the crisis. Although there has been an increase in the share of non-performing loans in the banking sector, this was well below the levels of many other large OECD economies. Banks have also implemented timely re-capitalisation to cope with possible funding strains. Unlike many countries, Australia's leading banks have managed to keep profitability close to pre-crisis levels and have maintained a strong financial position (OECD, 2012).

10. Excluding a sharp downfall in the drought-affected year 2007, per farm net capital additions across all rural industries increased to AUD 47 964 per year between 2000 and 2013, from AUD 20 435 per year between 1990 and 1999 (ABARES, 2014). Although this reflects the rise in land values and the falling number of farms, it may also suggest the stronger reliance of producers on borrowings to support investments.

11. On the ease of obtaining credit, respondents were asked: "In your country, how easy is it to obtain a bank loan with only a good business plan and no collateral?".

12. The World Bank's *Doing Business* estimate is based on the index of the strength of legal rights, depth of credit information (scope and accessibility of credit information distributed by public credit registries and private credit bureaus), coverage of individuals and firms by public credit registry, and coverage of individuals and firms by private credit bureaus.

13. Tax wedge on wage is the he combined central and sub-central government income tax plus employee and employer social security contribution taxes, as a percentage of labour costs defined as gross wage earnings plus employer social security contributions; the tax wedge includes cash transfers (OECD Tax Database).

14. Higher tax rates are applied in the United States, Japan, France, Belgium, Portugal and Germany. The new Australian government committed to reduce corporate income tax rate to 28.5%.

15. A Minerals Resource Rent Tax (MRRT) was introduced in July 2012 replacing the previous system of resource royalties. The mining industry stakeholders campaigned against the MRRT. The tax was levied on 30% of the "super profits" from the mining of iron ore and coal. Part of the proceeds from the MRRT was intended to enhance tax incentives for investment and productivity across broader business, with particular focus on SMEs. An important move in this direction was the introduction of a loss-carry-back arrangement applicable to corporate profits, which has been eliminated with the repeal of the MRRT in 2014. Another significant development in 2014 affecting the incentives of the resource sectors was the repeal of the carbon cap scheme introduced in 2012 (Box 5.2).

References

ABARES (2014), AGSURF Farm Survey on-line database, http://apps.daff.gov.au/AGSURF/.

ABS (2014), Innovation in Australia Business 2012/13, cat. No. 81580, Australian Bureau of Statistics, August. http://www.abs.gov.au/ausstats/abs@.nsf/mf/8158.0

AFMA (2013), *2013 Australian Financial Markets Report*, Australian Financial Markets Association https://www.afma.com.au/afmawr/_assets/main/lib90013/2013%20afmr.pdf.

Australian Treasury (2014), *The 2013 Tax Expenditure Statement*, the Treasury, Commonwealth of Australia, January 2014, Canberra.

Australian Treasury (2010), *Australia's Future tax System: Final Report*, the Treasury, Commonwealth of Australia, January 2010, Canberra. http://taxreview.treasury.gov.au/Content/Content.aspx?doc=html/home.htm.

Paul Conway, et al. (2007), "Product market regulation and productivity convergence", *OECD Economic Studies*, Vol. 2006/2. DOI: http://dx.doi.org/10.1787/eco_studies-v2006-art9-en

FIRB (2014), *Foreign Investment Review Board: Annual Report 2012-13*; Government of Australia, Foreign Investment Review Board, www.firb.gov.au.

Gibbs, C., K. Harris-Adams and A. Davidson (2013), *Review of Selected Regulatory Burdens on Agriculture and Forestry Businesses*, report prepared for the Department of Agriculture Agricultural Productivity Division, ABARES, Canberra, November.

Golub S.S. et al. (2003), "The Influence of Policies on Trade and Foreign Direct Investment", *OECD Economic Studies*, Vol. 2003/1. DOI: http://dx.doi.org/10.1787/eco_studies-v2003-art2-en

Gray, E.M., Oss-Emer, M. and Sheng, Y. (2014), *Australian Agricultural Productivity Growth: Past Reforms and Future Opportunities*, Australian Bureau of Agricultural and Resource Economics and Sciences Research Report 14.2, Canberra, February, http://www.daff.gov.au/abares/publications .

Havarnek, T. and Z. Irsova (2012), Survey article: Publication Bias in the Literature of Foreign Direct Investment Spillovers", *The Journal of Development Studies*, Vol. 48, No. 10.

Moir, B. (2011), *Foreign Investment and Australian Agriculture*, Rural Industries Research and Development Corporation, Canberra, November 2011.

NWC (2011), *Strengthening Australia's Water Markets*, National Water Commission, Canberra http://nwc.gov.au/__data/assets/pdf_file/0008/11240/StrengtheningAustraliasWaterMarketsReport.pdf.

OECD (2014), *OECD Stat on-line database*, Trade Indicators, http://dotstat.oecd.org/Index.aspx.

OECD (2013), *National Treatment for Foreign-Controlled Enterprises: Including Adhering Country Exceptions to national Treatment 2013*, http://www.oecd.org/daf/inv/investment-policy/national-treatment-instrument-english.pdf.

OECD (2012), *OECD Economic Surveys: Australia 2012*, OECD Publishing, Paris. DOI: http://dx.doi.org/10.1787/eco_surveys-aus-2012-en.

OECD (2010a), *Measuring Innovation: A New Perspective,* OECD Publishing, Paris. DOI: http://dx.doi.org/10.1787/9789264059474-en.

OECD (2010b), *OECD Reviews of Regulatory Reform: Australia 2010: Towards a Seamless National Economy,* OECD Publishing, Paris. DOI: http://dx.doi.org/10.1787/9789264067189-en.

OECD (2010c), *The OECD Innovation Strategy: Getting a Head Start on Tomorrow*, OECD Publishing, Paris.
DOI: http://dx.doi.org/10.1787/9789264083479-en.

OECD (2008), *OECD Environmental Performance Reviews: Australia 2007*, OECD Publishing, Paris.
DOI: http://dx.doi.org/10.1787/9789264039612-en .

OECD (2007), "Product Market Regulation and Productivity Convergence", Chapter 5 in OECD (2007), *Economic Policy Reforms 2007: Going for Growth*, OECD Publishing, Paris.
DOI: http://dx.doi.org/10.1787/growth-2007-en.

Productivity Commission (2005), *Review of the National Competition Policy Reforms*, Report No. 33, Canberra.

Productivity Commission (2007), *Annual Review of Regulatory Burdens on Business: Primary Sector*, Productivity Commission, Canberra.

Productivity Commission (2014), *Submission to the Agricultural Competitiveness Taskforce*, April 2014, Canberra.

Sanyal, K. (2014), "Foreign Investment in Australian Agriculture", *Research Paper Series, 2013-14*, Parliament of Australia, Canberra.

WDI (2013), *World Development Indicators Database*. World Bank, http://data.worldbank.org/data-catalog/world-development-indicators.

World Bank (2013), *Doing Business 2014*, World Bank and the International Finance Corporation, World Bank Publications, Washington, DC. http://www.doingbusiness.org/reports/global-reports/doing-business-2014.

World Bank Group and PwC (2013), *Paying taxes 2014 – The Global Picture*, PwC, World Bank and IFC, Washington, DC. http://www.pwc.com/payingtaxes.

World Economic Forum (2013), *The Global Competitiveness Report 2013-2014*: Full data Edition, Geneva 2013. http://reports.weforum.org/the-global-competitiveness-report-2013-2014/#=.

WTO (2011), *Trade Policy Review: Australia—Report by the Secretariat (Revision)*, Trade Policy Review Body, WT/TPR/S/244/Rev.1, 18 May, World Trade Organization, Geneva.

WTO (2014), *Trade Profiles, On-line Statistics Database*. http://stat.wto.org/CountryProfile/WSDBCountryPFHome.aspx?Language=E.

Annex 4.A1

Australia's National Competition Policy Programme (1995-2004): Competition policy reform

In 1996, each Australian jurisdiction examined their entire stock of laws for potential restrictions on competition and identified 1 800 pieces of legislation for review. By 2004, nearly three-quarters of priority areas and 90% of non-priority areas had been completed. The guiding legislative principles and the legislation were reformed accordingly.

General reforms

- Extension of the anti-competitive conduct provisions in the Trade Practices Act (1974) to unincorporated enterprises and government businesses.
- Reforms to public monopolies and other government businesses:
  - structural reforms — including separating regulatory from commercial functions; and reviewing the merits of separating natural monopoly from potentially contestable service elements; and/or separating contestable elements into smaller independent businesses; and
  - competitive neutrality requirements involving the adoption of corporatised governance structures for significant government enterprises; the imposition of similar commercial and regulatory obligations to those faced by competing private businesses (such as liability for taxes or tax equivalent payments, dividends and rate of return requirements); and the establishment of independent mechanisms for handling complaints that these requirements have been breached.
- Independent authorities to set, administer or oversee prices for monopoly service providers.
- Third-party access on reasonable terms and conditions to essential infrastructure services with natural monopoly characteristics, on reasonable terms and conditions, under a general national regulatory regime.
- Review of legislation to assess whether regulatory restrictions on competition are in the public interest and, if not, what changes are required. Legislation reviewed has dealt professions and occupations, statutory marketing of agricultural products, fishing and forestry, retail trading, transport, communications, insurance and superannuation, childcare, gambling; and planning and development services.

Sectoral reforms

- Electricity: Structural, governance, regulatory and pricing reforms to introduce greater competition into electricity generation and retailing and to establish a National Electricity Market in the eastern states.
- Gas: A similar suite of reforms to facilitate more competitive supply arrangements and to promote greater competition at the retail level.
- Road transport: Implementation of heavy-vehicle charges and a uniform approach to regulating heavy vehicles to improve the efficiency of the road freight sector, enhance road safety and reduce the transactions costs of regulation.
- Water: Institutional, pricing and investment measures to achieve a more efficient and sustainable water sector, and implementation of arrangements that allow for the permanent trading of water allocations.

Source: OECD (2010b) based on Productivity Commission (2005).

Annex 4.A2

Agriculture value chain and the impact of regulations

Key Australian Government involvement/regulation	Key stages of agricultural cycle	Key state/territory government involvement/regulation
• Aboriginal land rights/native title • Environmental protection and biodiversity conservation	Acquisition of arable land	• land use and planning regulation • Aboriginal land rights/native title
• Aboriginal and Torres Strait Islander cultural heritage • Natural heritage, world heritage • International treaties and conventions covering natural and cultural heritage • Licensing and approval of chemicals, fertilizers and pesticides • Environmental protection and biodiversity conservation	Preparation of land	• Land use and planning regulation • Native vegetation legislation • Water regulation • Weed and vermin control regulation • Laws relating to Aboriginal and Torres Strait Islander cultural heritage, archaeological and Aboriginal relics, sacred sites • Use of chemicals, fertilizers and pesticides • Natural heritage • Environmental protection/assessment • Building regulations
• Chemical and pesticide supply and registration • Access to drought support • Fuel tax regulation • National pollutant inventory • Biosecurity regulation • Immigration regulation • Water access and regulation • Research and development funding and support	Farming • cropping • animal husbandry	• Animal welfare regulation • Transport regulation affecting use of farm machinery • Vehicle and machinery licensing regulation • Livestock regulation and identification • Access to drought support • Workplace, health and safety regulation • Fire control regulation • Weed and vermin control regulation • Livestock disease control regulation • Livestock movement regulation • Water access and regulation • Chemical and pesticide use
• Export certificates • Industrial relation regulations • Immigration regulation • Environmental regulation • Industrial relations regulation • National pollutant inventory	On-farm processing	• Building regulations • Machinery operations • Certification and labelling • Industrial relations regulation • Workplace health and safety regulation
• National land transport regulatory frameworks • Shipping and maritime safety laws • International maritime codes and conventions • Competition laws/access regimes • Animal welfare	Transport and logistics	• Transport regulations • Government-owned public/private transport infrastructure • Access regimes

Key Australian Government involvement/regulation	Key stages of agricultural cycle	Key state/territory government involvement/regulation
• Marketing legislation (mandatory codes and acquisition) • Food safety regulation • Quarantine regulation • Export controls • Export incentives • WTO obligations • Market access and trade agreements • Taxation	Marketing • Boards • Customers	• Interstate certification arrangements • Taxation

Source: Gibbs, C., K. Harris-Adams, and A. Davidson (2013), *Review of Selected Regulatory Burdens on Agriculture and Forestry Businesses*, report prepared for the Department of Agriculture Agricultural Productivity Division, ABARES, Canberra, November.

Chapter 5

Capacity building and services for the Australian food and agriculture system

This chapter outlines the role of infrastructure capacity, skills and education in facilitating innovation in agri-food. It describes the governance of policies to improve rural infrastructure, outlines main regional programmes and reviews briefly the quality and coverage of rural services. It then discusses efforts to respond to skills demand from the agri-food sector through labour, immigration and education policy. It also reports on trends in education expenditure and outlines the performance of education system. Finally, it provides an overview of education levels in agricultural, and enrolment in agricultural programmes, outlining the gap between skills supply and demand in the sector.

The statistical data for Israel are supplied by and under the responsibility of the relevant Israeli authorities. The use of such data by the OECD is without prejudice to the status of the Golan Heights, East Jerusalem and Israeli settlements in the West Bank under the terms of international law.

Infrastructure and rural development policies

Infrastructure is a linking element of the economic system that enables the movement of factors of production, goods, and information across agents and markets. As such, infrastructure is important in determining the location of an economic activity and the kinds of activities or sectors that can develop within the economy. The availability and quality of infrastructure affect the decisions of firms and individuals to invest. Information and communication technologies (ICT) have a distinct role in increasing innovation by speeding up the diffusion of information, facilitating networking, reducing geographic limitations, and improving efficiency in communication. OECD analysis at the firm level shows that the probability to innovate increases with the intensity of ICT use. This is true for both manufacturing and service firms and for different types of innovation. Today, high-speed broadband networks support innovation as much as electricity and transport networks spurred innovation in the past; they dramatically change practices, extend innovation process beyond firms, government and researchers to consumers. Internet access is a prerequisite for using e-government services, and the data indicate a strong co-relation between penetration of the broadband infrastructure and the use of e-government services by citizens. The importance of ICT for innovation also comes from the fact that it improves basic competences of people and thus their capacity to create knowledge and uptake new products. For example, OECD student performance assessment shows that years of computer use are associated with higher student proficiency levels in science (OECD, 2010a).

All levels of government in Australia are involved in providing infrastructure: state and territory governments are responsible for most types of public infrastructure, notably, electricity, ports, rural roads, and water supply. The Australian government is the provider of aviation and telecommunication services, and shares responsibilities with state and territory governments for national roads and non-urban rail systems (Productivity Commission, 2014).

There has seen significant infrastructure development in Australia over the past decade and a half, but the mining boom placed the spotlight on infrastructure capacity constraints. Growth in mineral exports exposed substantial strain on key port infrastructure and gaps in rail infrastructure. A high proportion of the electricity infrastructure needs replacement and the transmission network faces capacity constraints (OECD, 2010b). As Figure 5.1 shows, Australia lags substantially behind its principal competitors in agro-food trade by the road and railway density and has less intensive port traffic (although such cross-country comparison should be made with care given the differences in the countries' geographical conditions, and economic and business context). Businesses in Australia evaluate the quality of the transport infrastructure somewhat less favourably than in OECD countries overall and substantially below businesses in the five top performing OECD countries (Figure 5.2). Australia is roughly at par with its agro-food trade competitors by ICT penetration and local businesses appreciate the quality of the associated infrastructure (Figures 5.3 and 5.4). Nevertheless, access to broadband Internet in Australia is more limited and this service has slow connection speeds and is higher priced relative to some of the best internet performers, such as Korea or Japan (OECD, 2010b).

The audits by Infrastructure Australia established that priorities for infrastructure development concern the rail freight network, ports, energy market, water supply, and urban transport. The infrastructure challenge is increased by the need to support future economic development and to address environmental and climate change concerns. The infrastructure deficit is currently estimated at AUD 300 billion (Infrastructure Australia, 2013).

Since the late 1980s, and particularly with the adoption of the National Competition Policy (1995), Australia's infrastructure sector has undergone important restructuring and opening to competition. By the mid-2000s, half of the sector's capital investment had been undertaken by private companies, while prior to these reforms it accounted for about one-third of infrastructure investment (Nguyen et al., 2013). Capacity problems encountered during the mining boom suggest management difficulties in dealing with market failures, as well as gaps in policies. For example, the *2010 OECD Economic Survey of Australia* highlighted weak co-ordination between public infrastructure development and fiscal management, and the lack of co-ordination between the various levels of government and between jurisdictions of the same level. Other policy issues affecting investment decisions concerned pricing of public infrastructure (e.g. water and road use charges), complicated regulatory framework for private investment, and access to infrastructure services (OECD, 2010b).

In late 2010, the various Australian Governments intensified infrastructure reforms to address the emerging challenges. These reforms were largely focussed on improving co-ordination, planning and finance options for infrastructure projects, as well as on improving regulations to optimise the use of existing infrastructure and investment decisions. Public spending on infrastructure received a substantial boost through new financial arrangements and (co)financing initiatives (Box 5.1). At present, progress has been made in many directions, but the most recent policy assessments point to important issues that remain to be tackled (OECD, 2012b; Productivity Commission, 2014).

Figure 5.1. Australia's availability of transport infrastructure, international comparison, 2012

Normalised to 1 for the value of Australia

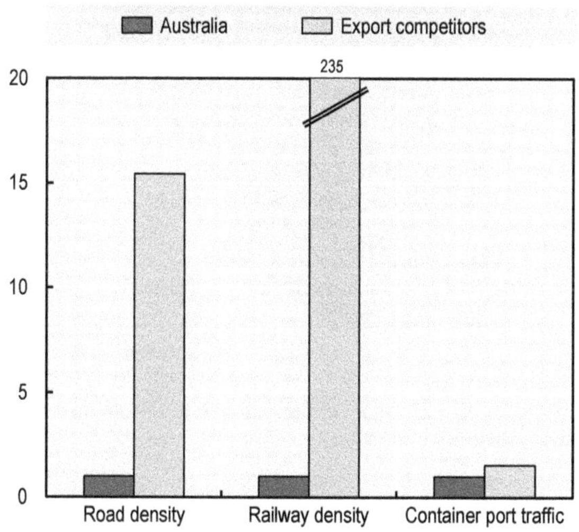

Export competitors are top 10 world exporters relative to Australia's leading agro-food exports (Argentina, Belgium, Canada, France, Germany, Ireland, Netherlands, New Zealand, United Kingdom and the United States). Data for road density refers to 2011.

Source: WDI (2013), *World Development Indicators Database*. World Bank, http://data.worldbank.org/data-catalog/world-development-indicators.

StatLink http://dx.doi.org/10.1787/888933249914

Figure 5.2. Global Competitiveness Index: Index of quality of Australia's transport infrastructure, 2013-14

Scale 1 to 7 (best)

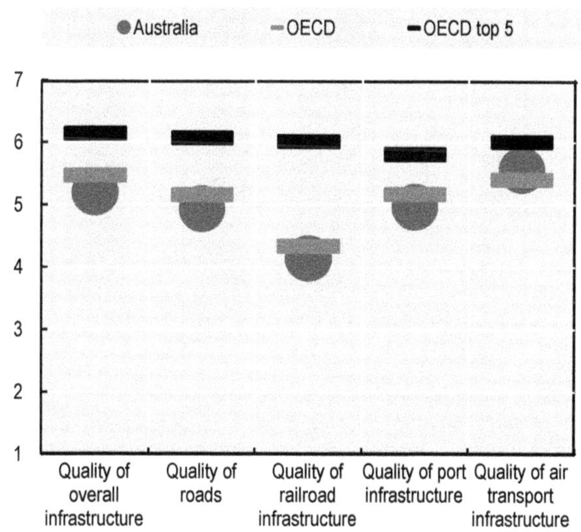

OECD top 5 refers to the average of the scores for France, Germany, Spain, Netherlands and Japan on the aggregate index of transport infrastructure quality.

Source: World Economic Forum (2013), *The Global Competitiveness Report 2013-2014: Full data Edition*, Geneva 2013. http://reports.weforum.org/the-global-competitiveness-report-2013-2014/#.

StatLink http://dx.doi.org/10.1787/888933249921

Figure 5.3. Australia's ICT penetration, international comparison, 2012

Normalised to 1 for the value of Australia

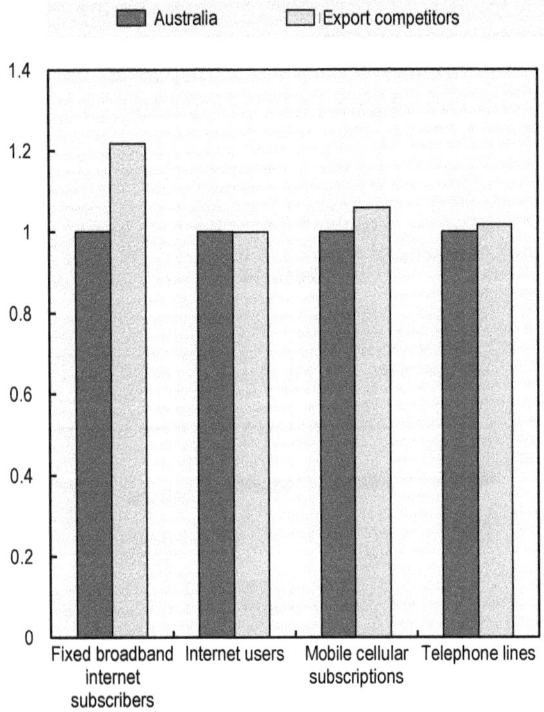

Export competitors are top 10 world exporters relative to relative to Australia's leading agro-food exports (Argentina, Belgium, Canada, France, Germany, Ireland, Netherlands, New Zealand, United Kingdom and the United States).

Source: WDI (2013), *World Development* Indicators *Database*. World Bank, http://data.worldbank.org/data-catalog/world-development-indicators.

StatLink http://dx.doi.org/10.1787/888933249939

Figure 5.4. Global Competitiveness Index: Index of quality of Australia's telephony and electricity infrastructure, 2013-14

Scale 1 to 7 (best)

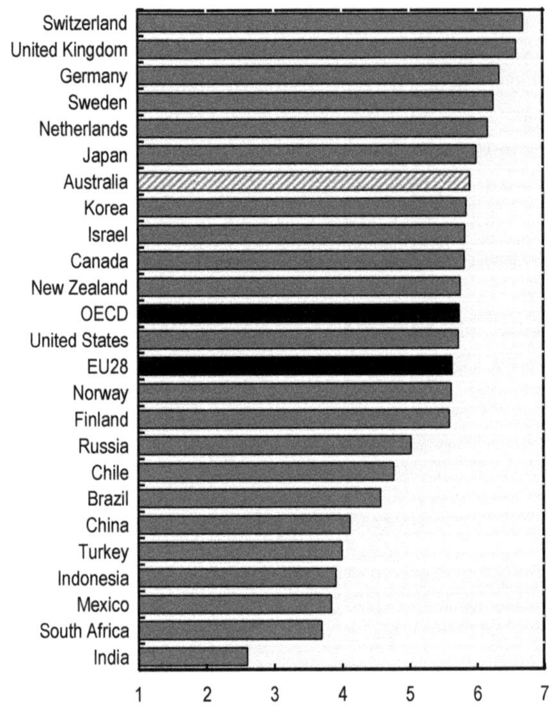

Indexes for EU-28 and OECD represent simple averages of member-country indexes.

Source: World Economic Forum (2013), *The Global Competitiveness Report 2013-2014: Full data Edition*, Geneva 2013. http://reports.weforum.org/the-global-competitiveness-report-2013-2014/#.

StatLink http://dx.doi.org/10.1787/888933249946

Box 5.1. Recent reforms of the infrastructure sector in Australia

Infrastructure sector reforms since the late 2000s have covered virtually all aspects of sector's operations – planning and funding, pricing, regulations, and institutions – and key infrastructure sectors.

Infrastructure planning and funding

Commonwealth government increased its involvement in the infrastructure sector. A federal body, Infrastructure Australia, was created in 2008 to support national focus on long-term strategic development of infrastructure. Infrastructure Australia is mandated to identify infrastructure needs, propose best financing mechanisms, policy, pricing, and regulation. This institutional change enabled to improve assessment and selection process for infrastructure provision, but as highlighted in the most recent Productivity Commission review there is scope for improvement (Productivity Commission, 2014).

Spending for public infrastructure has been substantially boosted; the GDP share of engineering and construction work for the public sector was doubled between 2000 and 2012, reaching slightly more than 2% (Productivity Commission, 2014). This increase in overall spending was accompanied by an increased share of Commonwealth contribution, including in the areas which were previously reserved for the states and territories. Currently, around 18 national infrastructure financing frameworks operate with contributions by Commonwealth and state/territories. One large framework is the Nation Building Program, which, among other areas, finances road, rail and port initiatives across Australia. This programme also finances the upgrading of Western Australia's Grain Rail Line based on co-financing by the Commonwealth and the West Australian Government. Another large framework is the Regional Infrastructure Fund which is based on proceeds from the Minerals Resource Rent Tax (MRRT) to finance critical infrastructure needs in areas supporting the mining industry.

Increased capacity to finance infrastructure has been of greater policy interest in the context of budgetary consolidation. The 2013 National Infrastructure Plan proposed a series of reforms to address infrastructure challenges. These include the establishment of a single national infrastructure fund to eliminate overlapping of multiple investment sources; using government funds differently to spread available funding; recycling the government capital in mature assets into new infrastructure; and promoting user financing based on the principle "user pays – says" (Infrastructure Australia, 2013). In November 2013, the Australian government initiated an inquiry by the Productivity Commission into the cost structure, decision-making and implementation process of infrastructure projects. A particular interest of the inquiry was to identify alternative ways of financing infrastructure investments with a view to encouraging private participation (Productivity Commission, 2014).

Harmonisation and reducing complexity of regulations, improving institutional arrangements

Efforts were directed towards harmonisation of regulations across jurisdictions. For example, nationwide rules for public-private partnerships (PPPs) were established, potentially lowering transactions costs for participants, facilitating inflows of private capital into infrastructure development, and enhancing incentives for project efficiency. For transport regulations, three single national regulators were created in rail safety and investigation, maritime freight and the heavy goods vehicle sectors. This is aimed at better national coherence of regulations and at reducing the cost of businesses that arise from cross-jurisdictional variations in regulations. Regulatory improvements included measures to rationalise the National Access Regime for third-party access to infrastructure. However, according to the Productivity Commission, inter-jurisdictional coherence and harmonisation of infrastructure regulations remain aspects where further progress needs to be made, while "institutional and governance arrangements remain deficient and are a major contributor to poor outcomes" (Productivity Commission, 2014).

Pricing of infrastructure services

The importance of pricing of infrastructure services to ensure appropriate incentives for infrastructure development is well recognised. Important progress was made to rectify water under-pricing and moving towards full cost recovery, in particular through improvements of water rights markets in rural areas within the National Water Initiative. Still, over-allocation of water resources remains an issue in some areas, as does full cost recovery for many water systems. The 2012 OECD Economic Survey of Australia also identified transport as another sector requiring less distortive pricing. In particular, it recommended broadening the application of road user charges for heavy vehicles to shift away from less direct system of charges producing cross-subsidisation between various types of vehicles.

Reforms in specific infrastructure sectors

Port and freight transport where capacity constraints were particularly manifest received special attention with the view to advance towards a single nationally integrated system. This implied better co-ordination between governments of various levels, improved long-term planning and regulation, and increased investment. The National Ports Strategy was endorsed by COAG in 2012 to support co-ordinated planning of port system development by state and territory governments, to reduce the "red tape" in execution of port plans, and to lower landside transport costs through improved port efficiency. This agreement was incorporated into a broader National Land Freight Strategy (2013).

A substantial reform was initiated in the telecommunications sector. One objective was to add competition to a sector dominated by the historical operator, Telstra. This involved changes in the infrastructure access regime by third parties, with a replacement of a negotiation/arbitrage model for a system of advance setting of uniform conditions and price of access for all applicants. Most importantly, the reform sought to overhaul the market structure and to develop a National Broadband Network to replace the copper telecommunications network.

Box 5.1. Recent reforms of the infrastructure sector in Australia (*cont.*)

The energy market was also covered by reforms seeking to increase competition in retail markets and improve regulation of transmission networks, to improve the handling of increases in renewable energy.

Sources: OECD (2010b), *OECD Economic Surveys: Australia*; OECD (2012), *OECD Economic Surveys: Australia*; Productivity Commission (2014), Public Infrastructure, Draft Inquiry Report, Volume 1; Infrastructure Australia (2013), *National Infrastructure Plan June 2013*.

Adequate infrastructure is a prerequisite for the future growth of the agri-food sector by making it more cost-competitive and increasing its capacity to channel products along the domestic supply chain and overseas. In 2008/09, the latest year available, infrastructure services accounted for 11% of total intermediate input costs in the agriculture, forestry and fishing sectors, 10% in the food processing industry, and 14% in the food services industry (including storage facilities) (Nguyen et al., 2013 based on ABS data).

Box 5.2. Infrastructure demand for Australia's agro-food industry

A global economic simulation model developed by ABARES predicts a 77% increase in real value of world agro-food demand between 2007 and 2050, with over two-thirds of this increase expected to come from the Asian region. Proximity to these markets provides Australia a comparative advantage and growth of these markets will become a significant driver for the growth of its agro-food industry. Australia's agro-food production is estimated to expand in real value at 1.3% per year and agro-food exports at 2.1% per year up to 2050.

Production and export growth will concern all key agricultural commodities, but is likely to generate different infrastructure pressures across industries:

- Wheat: While current distribution networks would appear to be adequate to process future wheat production in years with average seasonal conditions, additional capacity is likely to be needed in good years.
- Sugar: To accommodate significant increases in production and exports of sugar, improvements in handling capacities may be needed along the supply chain.
- Beef and sheep meat: As beef and sheep meat production increases it will be important to address issues in the livestock supply chain, including network connectivity with major supply chain points for high productivity vehicles, the condition of roads, and the competition for access to ports.
- Dairy: Achieving substantial dairy production increases is likely to require substantial increases in the movement of fodder and milk in some regions which would contribute to pressure on regional road networks.

Infrastructure will be important in making the most of opportunities presented by growing Asian markets:

- The road network is of particular importance and it will become increasingly important to tackle issues affecting its performance, including connectivity across the road network and with other transport modes, road and rail pricing distortions, funding arrangements and the integration of planning and investment across modes;
- Additional needs arise from the potential expansion of production in areas currently without adequate infrastructure: for example, a significant expansion of beef production in northern Australia would need to be supported by large infrastructure investments; similarly, to achieve substantial increases in dairy production it is likely that additional on and off-farm investment in irrigation infrastructure will be needed.

Climate change adds other dimensions to agro-food infrastructural development:

- The spatial and seasonal patterns of agricultural production may alter, requiring additional infrastructure in new areas; more frequent and severe extreme events may affect construction specification and costs of new projects and of maintaining existing structures.
- Reducing GHG emissions requires investment in management of energy, rural water, and waste.

Finally, given that infrastructure serves multiple sectors, requirements for the agro-food industry also depend on overall demand for infrastructure and growth perspectives in the rest of the economy.

Sources: Nguyen, N. et al. (2013), Infrastructure in Australia's Food Industry: Preliminary Economic Assessment; Linehan, V. et al.(2012), Food Demand to 2050: Opportunities for Australian Agriculture.

The existing estimates of currently unmet and future infrastructure needs vary, and various sources note that any such estimates require better information and better understanding of future demand for infrastructure. Infrastructure development needs likely differ across key agro-food industries; they also depend on potential changes in spatial distribution of agriculture due to climate change and the need for this sector to contribute to environmental improvements (Box 5.2 above).

Labour market policy

Australia has a growing population, due largely to immigration, and an increasing labour force. The labour market has performed well in the recent decade as compared to most other OECD countries. Agriculture has a small and declining share of total employment, and is characterised by the oldest age profile across the Australian industries. There is an apparent mismatch between supply and demand of skills in agriculture, which is also observed in some food industry segments (Box 5.3).

Box 5.3. Demographic and labour market trends in Australia

The Australian population increased by 1.5% per year between 2000 and 2012, a rate that exceeded nearly all OECD countries. A substantial portion of this growth is due to immigration. In 2007-12, net migration accounted for around two-thirds of yearly population increases (ABS, 2013a). Mid-range projections up to 2051 indicate that Australia population will increase from its current level of nearly 24 million to 38 million (ABS, 2013c). Despite projected net migration, Australia's population is expected to be ageing, although currently the country has still a relatively low old-age dependency ratio compared to countries such as Italy, Germany and Japan (OECD, 2012a with reference to ABS). Australia's labour force is increasing in absolute and relative terms. Labour market performs relatively well: the employment rose from 58% of the civilian population in 1999 to 62% in 2012, although high employment rates are associated with a significant incidence of part-time work, which is not always voluntary. Over the long term, there has been a substantial fall in the number of industrial disputes and the number of working days lost, as well as in the incidence of serious claims accepted for workers' compensation (ABS, 2013b). Nevertheless, in the conditions of the global slow-down and the subsiding of the mining boom, some reversal of positive trends has been recently observed as compared to the late 2000s. For example, employment growth decelerated, while the proportion of the underemployed has increased (Australian Government, 2014a).

Agriculture employs 285 200 persons, or 2.4% of national employment (November 2013); an additional 200 000 persons work in the food processing industry (2010/11 data). Agriculture is a sector with falling employment where over a quarter work part-time. Agriculture has the oldest age profile of any industry with 23% aged 65 years and older compared to just 3% for all occupations. Agricultural labour is the fastest ageing economic sector in Australia: the median age of farmers increased by nine years between 1981 and 2011, while for other workers it increased by six years (ABS, 2012). Agricultural professions score modestly on the occupation list in terms of the level of earnings. There is evidence of a mismatch between demand and supply of skills. While the National Farmers Federation estimates a shortfall of people in the top seven types of agricultural jobs, some professions (e.g. agricultural technicians, agricultural forestry and horticultural operators, farm workers) currently experience high unemployment (Australian Government, 2014c). In 2013/14, only 75% of vacancies were filled in agriculture and horticulture occupations and for a total of ten applications per vacancy only two were suitable for the vacancy; for example, recruiting farm managers for supervisory roles is reportedly difficult. In food trades, only 51% of vacancies were filled in 2013/14 and for 11 applications per vacancy, less than two were considered suitable. Out of four assessed occupations in food trades, all have been experiencing skill shortages for at least the last 4-5 years (Australian Government, 2014c).

Labour regulations affect the cost and conditions of employing labour, and thus production choices of firms and their incentives to invest in new products and processes. In 2009, Australia adopted a new labour framework regulation, the Fair Work Act 2009, which has, however, remained under intense debate since then. In particular, there have been claims on the part of business associations that the provisions of the new Act impede productivity growth. The Fair Work Act underwent a post-implementation review in 2012 which assessed the Act favourably, but the new government set it for a further review, which, through a broad and independent examination is intended to identify future options to improve laws (Box 5.4).

Box 5.4. Australia's labour regulation framework in evolution

Australia' labour regulations are currently governed by the Fair Work Act 2009, which made the system more employee-friendly. The previous labour Act, for instance, favoured statutory individual contracts, which were subject to the "fairness test" rather than the previous "no disadvantage" test. Under the Fair Work Act 2009, individual statutory agreements are not provided for and bargaining generally must be conducted at the enterprise level; a "better off overall" test has been introduced and the results of the negotiations have to be approved by the regulator, the Fair Work Commission. At the same time, collective agreements must provide for "individual flexibility arrangements" between an employer and an employee, and which do not need to be lodged with the regulator and can be terminated by either party with relatively short notice. The 2009 Fair Work Act has also broadened the minimum safety net conditions and restored protection against unfair dismissals for employees in SMEs (OECD, 2012b).

The post-implementation review of the Fair Work Act carried out in 2012 found that this new framework seemed to be yielding good results in terms of balancing employer-employee interests. This review did not suggest any fundamental change to the legislation, but in response to stakeholder feedback proposed 53 amendments to encourage flexibility and equity within the system. These amendments were aimed, for example, to promote firm-specific agreements favourable to productivity, including improved operation of individual flexible arrangements; facilitate the negotiation of labour agreements for new enterprises (greenfields agreements), and reduce the risks of improper recourse to unfair dismissal procedures (OECD, 2012b). The previous government implemented a number of recommendations of the 2012 review and the current Australian Government has committed to address a number of other outstanding issues identified by this review.

Nevertheless, business continued to voice concerns that the Fair Work Act provisions do not provide employers with sufficient flexibility to seize emerging market and investment opportunities. For example, they referred to the overly broad range of matters that can be bargained over, insufficient flexibility in negotiation of labour agreements for new large investment projects, and constraining work and pay regulations in specific service sectors (Business Council of Australia, 2014). In March 2014, the new government announced a further review of the Fair Work Act, which is intended to be a comprehensive examination by the Productivity Commission of principal aspects of the labour regime, and the impact they have on the economy, productivity and employment. The main focus is whether the legislation works for both employers and employees and does not impede productivity and competitiveness. This review is also expected to identify future options to improve the laws.

Australian business opinions are reflected in overall less favourable perception of labour market efficiency compared to businesses across the OECD area, which is due to relatively low business perceptions on aspects such as regulatory flexibility of wage determination, hiring and firing practices and the perceived impact of the pay system on labour productivity (Figure 5.5). The degree to which these perceptions are also shared by the businesses in agriculture and agro-processing industries has remained unexplored in this review due to a lack of information.

Current labour regulations in Australia lead to a relatively high minimum wage level at above 50% of the median full-time wage in 2010-12. This places the country in the group of OECD countries with the highest relative level of minimum wages. In combination with existing benefits, minimum wages enable full-time workers to escape relative income poverty, and Australia has one of the lowest in-work poverty rates across OECD countries (OECD, 2010). A relatively high safety net is established against high average wage earnings in Australia, which ranks fourth among OECD countries by that level (USD 49 655 in 2012 expressed in constant PPP terms). The total tax wedge on wage (27%) is below the OECD median rate (Figure 3.8) and is the fourth lowest in the OECD area (after Mexico, New Zealand, and Korea).

Beyond the cost of labour and conditions for employing it, labour regulations also affect innovation by their impact on labour mobility and structural adjustment. Innovative firms engaged in changing technologies, processes, or business organisation, are likely to be particularly sensitive to adequate conditions for hiring and dismissing people, complemented by good systems of unemployment insurance, and support for job placement, skills training and continuous learning.

Figure 5.5. Global Competitiveness Index: Index of labour market efficiency, 2013-14

Scale 1 to 7 (best)

A. Total index of labour market efficiency, international comparison

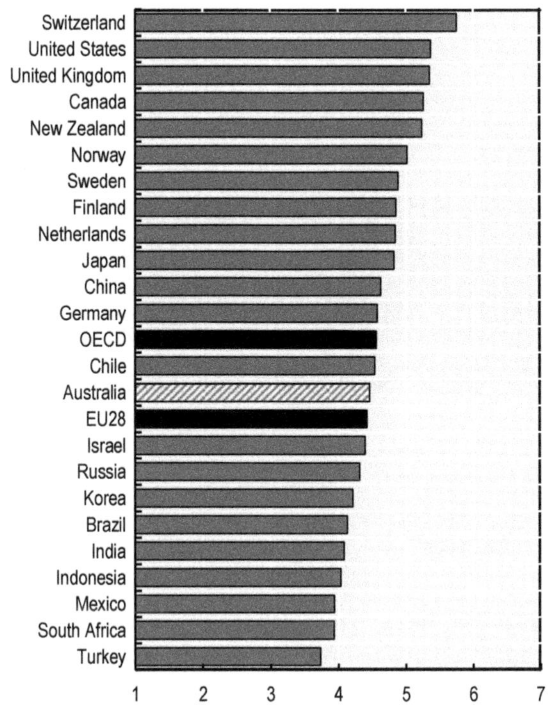

Indexes for EU-28 and OECD represent simple averages of member-country indexes.

B. Australia's index of labour market efficiency by component

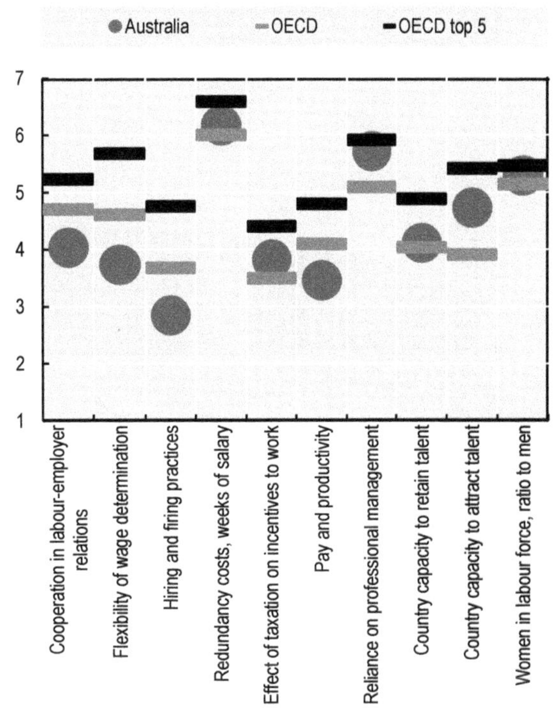

OECD top 5 refers to the average of the scores for the top 5 performers among OECD countries (Switzerland, United States, United Kingdom, Canada and New Zealand).

Source: World Economic Forum (2013), *The Global Competitiveness Report 2013-2014*: Full data Edition, Geneva 2013. http://reports.weforum.org/the-global-competitiveness-report-2013-2014/#.

StatLink http://dx.doi.org/10.1787/888933249955

Employment protection regulations directly affect labour mobility. The OECD indicators of employment protection legislation measure the procedures and costs involved in dismissing individuals or groups of workers, and the procedures involved in hiring workers on fixed-term or temporary work agency contracts. Australia's employment protection regulation has become stricter since the introduction of the Fair Work Act; the indicator of regular worker protection increased from 1.63 points in 2009 to 1.94 points in 2013, and from 0.79 points to 1.04 points for the temporary employed. Nevertheless, on an international scale, Australia's employment protection regulations are not very restrictive, both with regard to regular and temporary employment (Figure 5.5). Australia's labour market is characterised by a relatively high job turnover and low tenure among employees. For example, in 2012, 42% of the employed remained with the same employer for less than three years. A quarter of the employed had been with the same employer for ten years and longer; this compares to between 32% and 45% of workers in Canada, Germany, the Netherlands, and the United Kingdom (OECD, 2014b).

Figure 5.6. OECD indicators of employment protection legislation, 2013[1]

Scale from 0 (least) to 6 (most) restrictive

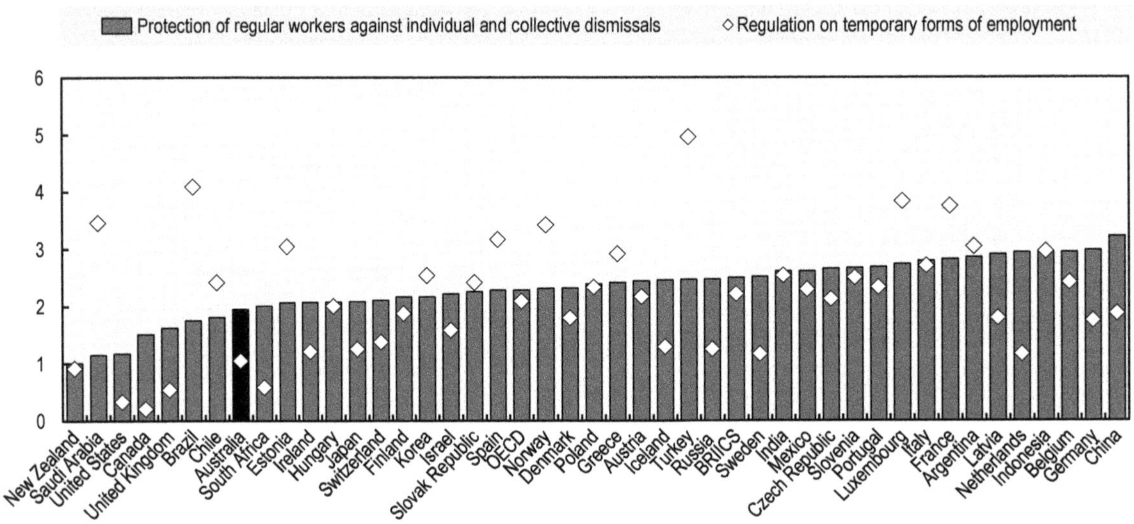

1. Data refer to 1 January 2013 for OECD countries and Latvia and 1 January 2012 for other countries.

Source: OECD Employment Protection Database, www.oecd.org/employment/protection.

StatLink http://dx.doi.org/10.1787/888933249967

A well-functioning employment programmes for those who have lost their jobs or face such a risk facilitate labour mobility. In the context of innovation, a balance between passive measures that provide temporary financial support to the unemployed and active measures oriented towards job placement, training, job rotation and sharing, employment and start-up incentives is of particular relevance. From this perspective, passive measures have precedence in Australia: in 2011, such measures accounted for over three-quarters of total public expenditures on labour market programmes and over two-thirds of participant's stock (OECD, 2014b).

However, active measures seem to gain in importance over time, which may in part be due to the employment services reform in 2009. Employment services in Australia, such as Job Services Australia, are offered through private providers who are federally funded in proportion to their success in returning the unemployed to work. Job Services Australia providers receive service fees and outcome payments for assisting job seekers to undertake an education course or gain employment. Providers may also receive a bonus payment where a job seeker undertakes an apprenticeship in a skills shortage area or where a job seeker gains employment directly related to an education qualification. For example, in 2014, 23% of the job seekers were engaged in education and training, of which the majority were studying at the certificate or diploma level, or higher, while 15% of those placed in jobs were in education three months after their job placement (Australian Department of Employment, 2014). Further changes are to be implemented as of July 2015 to increase incentives of employment providers to move job seekers off income support and into employment as efficiently as possible. Employment Providers also have access to resources and incentives for employers to assist job seekers, such as tailored training, targeted wage subsidies, purchase work related items and post-placement support to employers and job seekers.

Apart from activity of Job Services Australia, there are other job placement programmes and related services, with several targeted to or having relevance to agriculture; all of these programmes incorporate training components. These programmes consist of supporting self-employment through creation of SMEs, encouraging employment in remote areas, and facilitating responses to peak labour demand in agriculture (Box 5.5).

Box 5.5. Job programmes related to agricultural and regional employment in Australia

New Enterprise Incentive Scheme (NEIS): Provides job seekers with accredited small business training, business mentoring, and income support to help turn a business idea into a viable business, and to help the job seeker to become a self-employed business owner. Providers include local organisations, such as Business Enterprise Centres, TAFE Small Business Centres, community organisations, and the private sector businesses. Job seekers receive a NEIS Allowance for up to 39 weeks while they are operating their NEIS business. The amount received is not affected by the income received from the NEIS business and some participants may also be eligible to receive NEIS Rental Assistance.

Relocation Assistance to Take Up a Job: Assists eligible job seekers with the costs of relocating if they move to start a new job. Job seekers who move to a regional area can access up to AUD 6 000 of relocation assistance. Families with dependent children may be eligible for up to an extra AUD 3 000 to help cover the additional costs of relocation.

Tasmanian Jobs Programme: Encourages businesses to provide jobs to Tasmanians who have been unemployed for an extended period. It began in 2014 and is a two-year trial programme providing a one-off payment of AUD 3 250 to any Tasmanian business that employs eligible job seekers for a period of at least six months.

Harvest Labour Services: Australian job seekers and visitors to Australia who hold appropriate work visas can get help finding work picking fruit and other crops in rural and regional Harvest Trail areas across Australia. This programme only services short term and largely unskilled labour requirements of horticultural growers at harvest times. Over recent years there have been no critical labour shortages due to the numbers of overseas backpackers.

Drought Force: Provides job seekers with work experience and training in farming communities affected by drought. Jobs are undertaken on a voluntary basis only. Participants in Drought Force activities are offered structured training, where appropriate and available. This can include accredited and non-accredited training.

Remote Jobs and Communities Program (RJCP): Provides jobs, participation and community-development services in 60 remote regions across Australia. It aims to help build skills, obtain jobs for the unemployed and integrate people in community development. It also helps remote-area employers to meet their workforce needs.

Regional Education, Skills and Jobs initiative: Helps to increase awareness of communities in regional Australia and their participation in federal education, training and employment programmes.

Source: Australian Government, Department of Employment, https://employment.gov.au/employment.

Policy on international mobility of human resources, particularly of skilled people, affects the creation, diffusion, and uptake of knowledge and, as such, operates as an enabler of innovation. Immigration, the principal source of population growth in Australia, is also an important contributor to labour supply. Australia implements a policy which favours skilled migration linked to labour market needs and the majority of immigrants are skilled and highly skilled persons. For example, between January 2013 and March 2014, of the total number of persons who obtained permanent entry visa or permanently settled in Australia during that period, almost two-thirds came under the various schemes designed to attract skilled immigrants (including business innovation and investment programme; regionally-sponsored migration scheme; employer nomination scheme; and general skilled migration) (Australian Government, 2014d). There is also an important Worker Holiday Visa Programme, but it attracts people from overseas for short-term periods, particularly during seasonal peaks in horticulture.

According to the Australian government, overseas workers constitute a "small but important part of the agricultural industry" (Australian Government, 2014b). Agricultural businesses are able to recruit unskilled workers through short-term seasonal schemes. The role of overseas labour stretches beyond just meeting seasonal labour deficits. In the conditions of an ageing agricultural population and the relatively low attractiveness of agricultural jobs, immigrant labour is viewed to become a progressively important source to meeting both the labour and skill supply gaps. Immigration policy has recently been focussed to encourage immigrant flows towards regional and rural areas, however so far they have been directing labour predominantly to non-agricultural occupations (Box 5.6).

Australian agriculture is characterised by the highest share of employed labour which is above retirement age, a labour structure which is also aging more rapidly than in the rest of the economy. There are suggestions that with many farmers nearing retirement, inter-generational transition may tighten labour supply with a loss of tacit knowledge and experience. There is competition for scarce labour with more lucrative and dynamic sectors, and most rural areas struggle to attract the young and to retain workers (AgriFood Skills Australia, 2014). Industry groups feel uncertainty about the future labour supply in the agro-food industries and raise concerns about an apparent mismatch between demand for and supply of required skills (AgriFood Skills Australia, 2011; AgriFood Skills Australia, 2014; NFF, 2011; NSWG, 2012). Alternative views are that the substitution of capital for labour puts downward pressure on labour demand in agriculture and that the progressive reliance of farm businesses on external providers of knowledge services may compensate for the possible skills gap among farmers. Any assessment notwithstanding, the employment services system has a role to play in transmitting employer demand for agro-food skills to the education and training systems. Employment service also has a more direct role in contributing to filling the skills gap through job placement and training programmes. Australia's existing general employment services programmes may not be sufficiently focused on the specific needs of the agro-food industry, while current agriculture-specific labour programmes are only narrowly targeted to respond to temporary labour demand peaks. There may be scope for job placement programmes focused specifically on the agro-food industries and which respond to longer-term labour and skill needs of the sector. Similarly, there may be scope for immigration schemes that focus on agro-food industries. There may also be scope for greater integration within and between the labour services and immigration schemes to better cater to the needs of agro-food industries.

Box 5.6. Migration schemes related to regional employment in Australia

Two immigration schemes are in operation for seasonal unskilled labour.

- Seasonal Worker Program: Helps Australian employers in the horticulture industry to employ workers from eight selected Pacific Island countries and Timor-Leste when they cannot find sufficient local labour to satisfy seasonal demand. The Add-on Skills Training component allows seasonal workers to access basic training opportunities to increase their performance capability both in Australia and in their home country. Developing skills is an important feature and seasonal workers can receive formal qualifications under the Australian Quality Training Framework at no cost to the worker or the Approved Employer. This programme is relatively new, starting in 2012, and small: just over 2000 visas were issued in 2013/14, four times less than New Zealand's Recognised Seasonal Employer Scheme.
- Working Holiday Maker Programme is substantial, with almost 230 000 visa grants in 2013/14 alone. As the programme is targeted at young people, much of the labour coming through the Programme has very limited employment experience and is at the unskilled end of the spectrum. Furthermore, the areas where programme participants do work is not well aligned with the areas experiencing labour shortages (Tan et al., 2009).

Two longer-term schemes cater to needs for skilled and semi-skilled labour in regional Australia:

- Regional Skilled Migration Scheme (RSMS) provides for employer sponsorship for at least two years in a full-time skilled job. According to Collins et al. (2014), this policy has been quite successful, with new skilled immigrants generally filling labour shortages and adding to the productivity of the regional and rural economy. However, this scheme is not limited to agricultural workers – the top five occupations are cooks, restaurant managers, retail managers, chefs, and motor mechanics. Also, as the definition of "regional Australia" is very broad, under current arrangements a sponsoring employer can be located anywhere in Australia, except a few specified areas.
- Designated Areas Migration Agreements (DAMA) The purpose of the DAMA is to supplement the workforce strategies of states, territories and regions to support economic performance and help them adjust to changing economic conditions. Under the DAMA eligible employers can sponsor skilled and semi-skilled workers.

Source: Australian Department of Immigration.

Education and skills policy

Education policy has strong and diverse links to innovation. A high level of general and scientific education facilitates acceptance of innovations by society at large. Effective innovation systems require well-educated researchers, teachers, extension officers and business. Producers with a good general, technical and business education would generally be more willing and better skilled at adopting innovations. For adoption, new technologies and products require understanding of background scientific principles, and the gains and risks involved, by the producer.

Australia has a modern and well performing education system, with higher education being the country's third largest export industry (Box 5.7). In 2011, Australia was a key destination for overseas education hosting more than 6% of the world's foreign students (OECD, 2013a).

Australia's population is well educated, with a share of those who have received higher education (41%) above the OECD average (33%) (Figure 5.7). Nevertheless, compared to some other OECD countries, for example, Canada and the United States, Australia is less advanced in terms of the structure of education attainments, having a smaller proportion of people with higher education and higher share of people with below upper secondary education. Education attainments of the employed in agriculture have much improved since mid-1990s, but remain below the rest of the economy – in 2013, the percentage of employed with post-secondary education reached 47% in agriculture compared to 69% across all economy (Australian Government, 2014e).

Figure 5.7. Education attainments of Australia's population, international comparison, 2012

Percentage distribution of population between 25 and 65 years of age

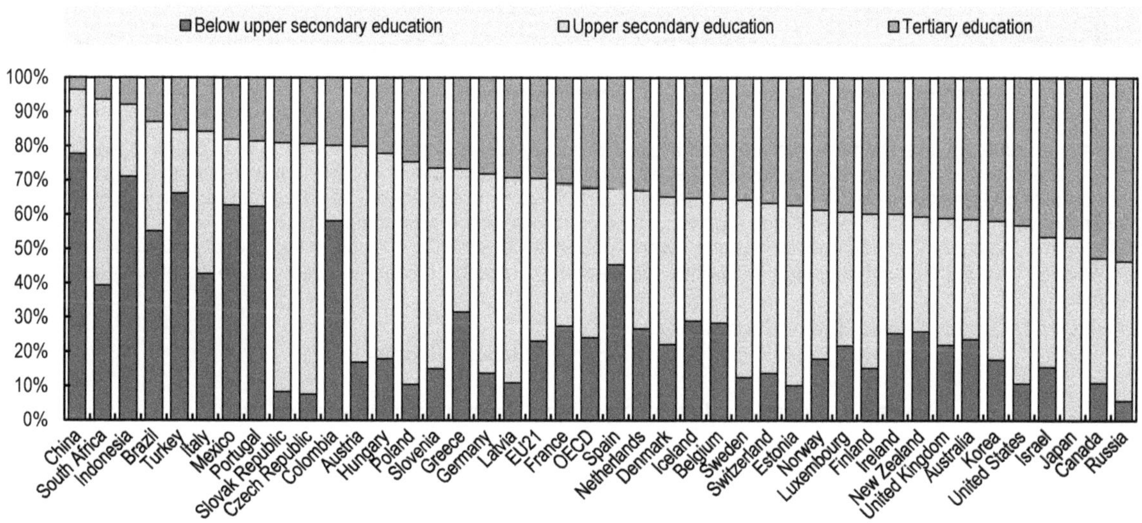

Data for Colombia and Indonesia represent 2011 and for China represent 2010. EU21 refers to EU countries member to the OECD.

Source: OECD (2014), *Education at a Glance*.

StatLink http://dx.doi.org/10.1787/888933249967

The OECD Programme for the International Assessment of Adult Competencies (PIAAC) assesses the performance of adults across 24 countries and sub-national regions in key information processing areas – literacy, numeracy and problem solving in technology reach environment. People aged 16-65 in Australia have above-average proficiency in literacy and problem solving in technology-rich environments compared with adults across other countries participating in the survey, and only an average proficiency in numeracy (Figure 5.8).[1] However, as in most countries surveyed, quite large proportions of adults in Australia demonstrate relatively low skills. Over a half of the surveyed scored below level 3 in numeracy (out of five levels), while in problem solving 62% of adults perform below level 2 (out of four levels) (Figure 5.9). Some 7.5% of Australian adults indicated that they had no prior experience with computers or lacked very basic computer skills in problem solving in technology-rich environments, while almost 14% opted out of the test. Still, the proportion of adults at lower proficiency levels in Australia is somewhat smaller than on average across surveyed countries, except in numeracy.

Figure 5.8. Shares of adults performing at different proficiency levels in literacy, numeracy and problem solving in Australia, 2012

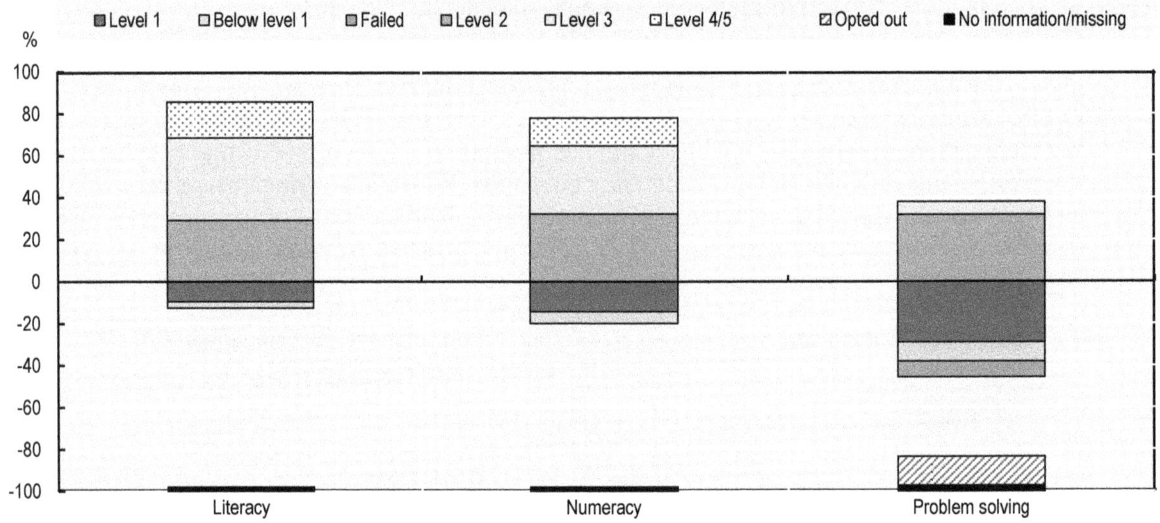

Numeracy and literacy assessments are reported in six proficiency levels (Levels 1 through 5, plus Below 1) while problem solving is reported in four proficiency levels (Levels 1 through 3, plus Below 1). Additional respondent groups to the problem-solving domain include those who 1) failed the ICT core assessment, 2) opted out of the computer based assessment version of the test, even if they reported some prior experience with computers.

Adults in the category "no information/missing" were not able to provide enough background information to impute proficiency scores because of language difficulties, or learning or mental disabilities (referred to as literacy-related non-response).

Source: OECD (2013c), *Survey of Adult Skills (PIAAC).*

StatLink http://dx.doi.org/10.1787/888933249980

Figure 5.9. Shares of adults performing at lower proficiency levels[1], 2012

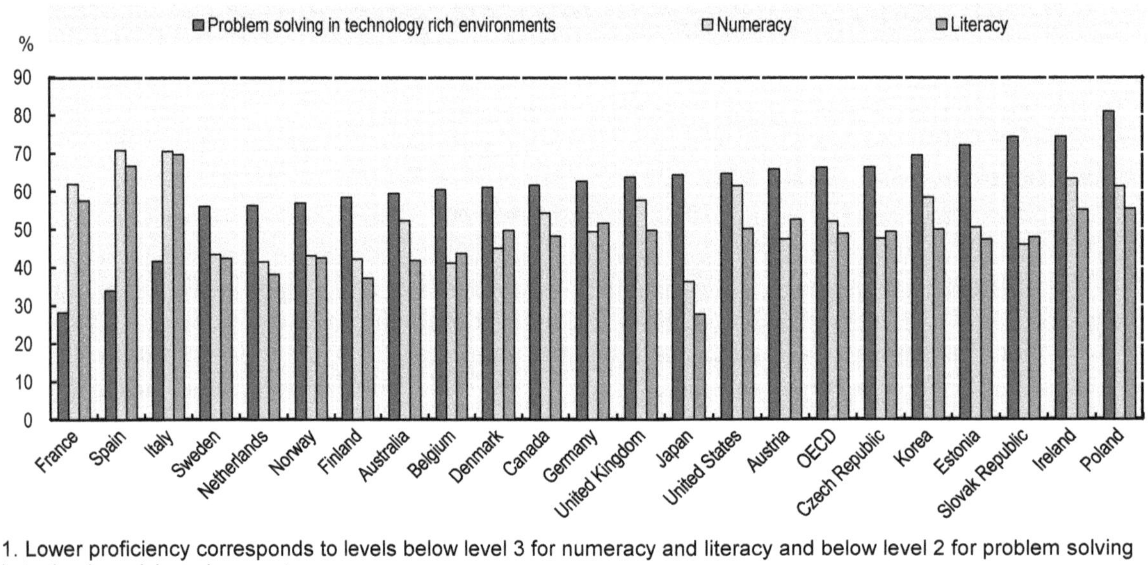

1. Lower proficiency corresponds to levels below level 3 for numeracy and literacy and below level 2 for problem solving in technology rich environment.

Source: OECD (2013), *Survey of Adult Skills (PIAAC).*

StatLink http://dx.doi.org/10.1787/888933249996

Business perceptions on various dimensions of Australia's education, such as quantity and quality of education, and on-job training, can be drawn from the WEF's Global Competitiveness Index component related to education and training. This index places the country above the average OECD level. This overall score reflects higher appreciation of Australian businesses of the quality and quantity of national education compared to average perceptions across OECD area. At the same time, the degree of appreciation of the national education system by businesses in Australia is markedly below five OECD countries with the highest aggregate indexes of education and training (Figure 5.10.A).

Figure 5.10. Global Competitiveness Index: Index of higher education and training, 2013-14

Scale 1 to 7 (best)

A. Total higher education and training index, international comparison

B. Australia's index of higher education and training by component

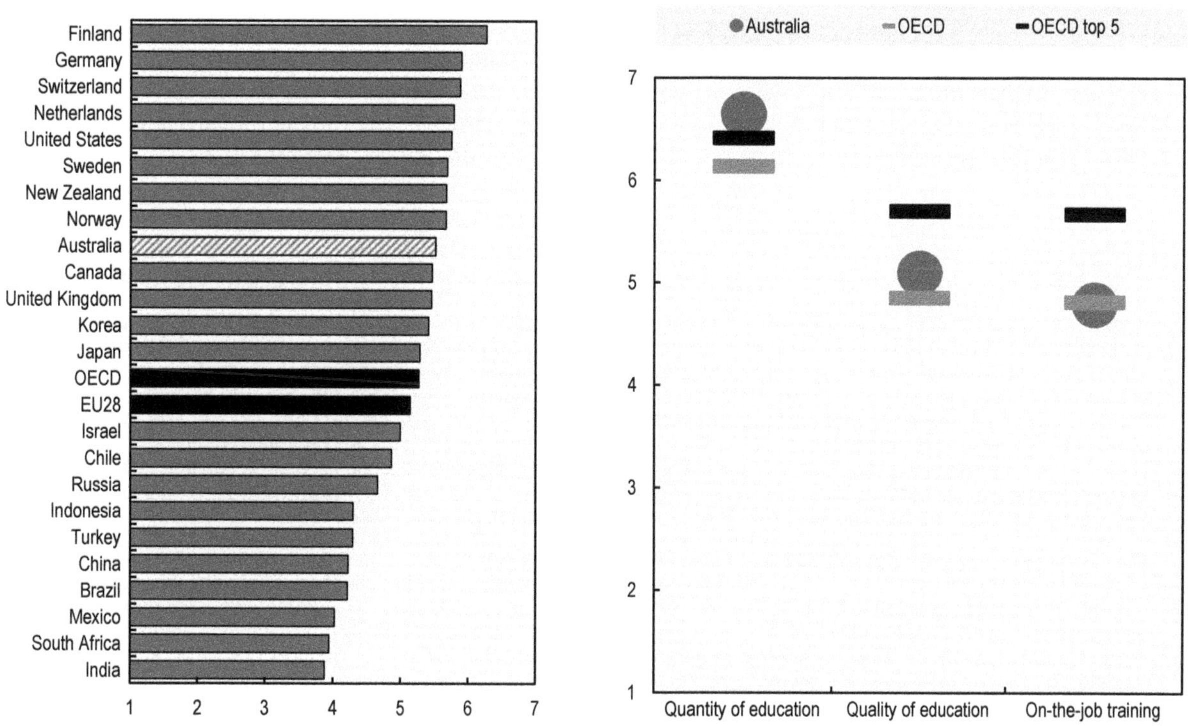

Indexes for EU-28 and OECD represent simple averages of member-country indexes.

OECD top 5 refers to the average of the scores for the top five performers among OECD countries – Finland, Germany, Switzerland, Belgium and the Netherlands.

Source: World Economic Forum (2013), *The Global Competitiveness Report 2013-2014: Full data Edition*, Geneva 2013. http://reports.weforum.org/the-global-competitiveness-report-2013-2014/#.

StatLink http://dx.doi.org/10.1787/10.1787/888933250005

Between 2000 and 2010, public expenditures on primary and secondary education increased by 58% and by 55% on tertiary education, resulting in higher portion of the national wealth directed to education. The increase in public spending on education in Australia was more significant than on average across OECD countries, still its relative level (5.9% of GDP) remains in Australia slightly below the OECD average level of 6.1% and further below OECD countries such as Canada, the United States, and New Zealand (Figure 5.11).

Australia is one of the leading OECD countries in student performance (Figure 5.12). According to OECD Programme for International Student Assessment (PISA), 15-year old students in rural and non-rural areas performed above average OECD students in key knowledge fields (mathematics, science and reading)[2]. Nevertheless, these results also show a gap with top five OECD performers, particularly in mathematics. 2012 PISA survey shows no progress in student performance in Australia in all three study fields since the previous survey in 2009. Rural 15-year olds represent an obvious pool of future workforce for the agri-food sector. PISA diagnoses a lagging performance of this group of students in Australia compared to their peers in non-rural areas, a feature common to the majority of surveyed countries. In Australia, the gap with top five OECD performers is more significant for rural students than it is for non-rural students. A lagging performance of rural 15-year olds is likely one of the factors leading to underrepresentation of rural population in tertiary education (see below).

Box 5.7. Australia's education system: Organisation and roles of governments at different levels

Australia's education is a decentralised but nationally steered system based on agreements between states and territories and the operation of national councils and national authorities focussed on specific areas of education. These arrangements are aimed to support development of a shared national education framework in terms of education priorities, standards, policies, and financing. The distribution of the roles between the state/territory and Australian government varies at different levels of education. State and territories have the leading role in schools and vocational education and training (VET). They make most planning, structure and resource decisions, whereas schools have substantial discretion in organisation of instruction. Decision-making in higher education is shared between the Australian government and higher education providers. While the Australian government assumes the policy responsibility, universities have the authority to accredit their courses and assure quality standards.

Public sources constitute almost three quarters of overall expenditure on educational institutions in Australia – this share is below the OECD average of almost 84% and Australia has the sixths largest share of private expenditures in the OECD area taking account of all levels of education. Funding contributions by governments of different levels vary depending on the level of education. In non-tertiary education, state and territory governments contribute almost 90% of public funding for public schools and somewhat less than 30% for private schools, with the rest of public funding provided by the Australian government. Public schools cater almost two thirds of Australian students who are entitled for free education, while one-third of students attend private schools and pay tuition. In higher education, public sources account for 46% of their total funding. The Australian government assumes the primary financing responsibility, with financing largely implemented in the form of subsidising tuition costs through the Commonwealth Grant Scheme.

Upper secondary education is not compulsory in Australia, and examinations vary across the states and territories. A challenge at this level is to provide relevant education that will prepare young adults for work and, at the same time, provide capacity for further learning. Australia has well developed system of vocational education and training (VET). It is provided at the general secondary and tertiary education levels, and employers are well-engaged in the system. VET can facilitate entry into the labour market through work-study programmes, Technical and Further Education (TAFE) institutes and private Registered Training Organisations (RTOs). In tertiary education, Australia has the sixth highest graduation rate among OECD countries in academic programmes (tertiary-type A), and the eighth highest graduation rate in vocationally oriented programmes (tertiary-type B). About one-fifth of all tertiary students are foreign, the highest rate of foreign tertiary students among OECD countries in 2011.

Source: OECD (2013b), *Education Policy Outlook: Australia*; *Agrifood Skills Australia* (2014), *2014 Environmental Scan of the Agrifood Industry.*

Figure 5.11. Australia's public and private expenditure on education as a percentage of GDP, international comparison, 2011

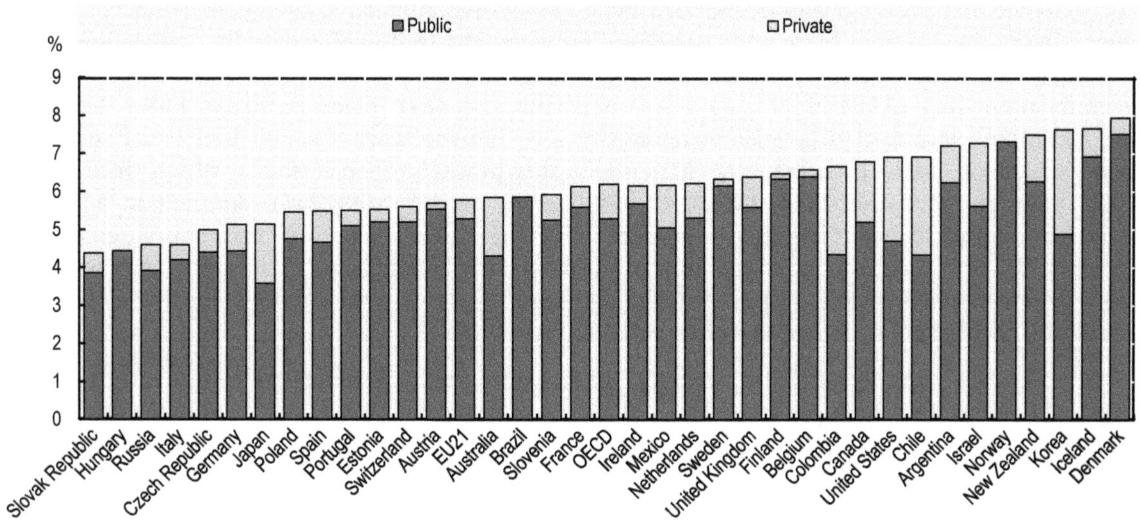

Public expenditure includes public subsidies to households attributable for educational institutions, and direct expenditure on educational institutions from international sources. Private expenditure is net of public subsidies attributable for educations institutions.

Data for Canada refer to 2010. Data for Chile refer to 2012

Source: OECD (2014a), *OECD Education at a Glance database*, 2014, http://www.oecd.org/education/eag.htm.

StatLink http://dx.doi.org/10.1787/10.1787/888933250016

Figure 5.12. PISA assessment 15-year old students' performance in mathematics, reading and science, 2012

Mean performance scores

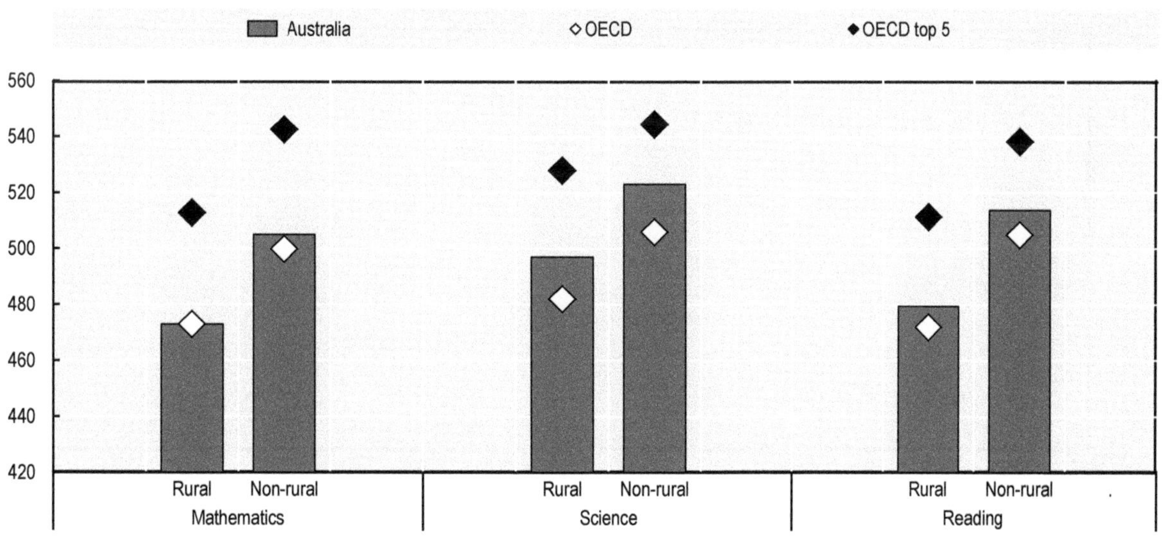

Rural students include students attending schools in villages and small towns; non-rural students include all other students. The categorisation of schools according to certain area is based on responses to a survey's question: "Which of the following definitions best describes the community in which your school is located: village, small town, town, city or large city?"

Source: OECD (2013d), *OECD Programme for International Student Assessment (PISA) database*, 2013, http://www.oecd.org/pisa/.

StatLink http://dx.doi.org/10.1787/888933250027

From the supply side, agricultural education in Australia is well funded and offers a variety of courses. Of a total of 39 universities in the country, twelve offer agricultural related degrees at the undergraduate and postgraduate levels, with most universities running related courses in business and science fields. In addition, there are agricultural colleges and technical and further education (TAFE) institutions with agricultural programmes. Agriculture units of study receive the highest rate of government funding (AUD 21 075 per Commonwealth supported place in 2013). From the demand side, agriculture and related courses at university level attracts only a small number of students. In tertiary education, students' demand for agricultural disciplines has been steadily falling since the early 2000s: between 1998 and 2012 the share of university students that have enrolled in agricultural disciplines almost halved, largely in parallel with the falling share of agricultural employment (Figure 5.13). This contrasts the trends in some leading OECD education performers, such as Finland, Japan, or Switzerland, where despite a declining share of agricultural employment the proportion of agricultural enrolments in higher education was stable or rising. 3

Among 29 disciplines, including general life sciences, that may be related to future work in areas of agriculture and food, biological sciences, environmental studies, food science and biotechnology, and veterinary science are the most popular. Notably, disciplines directly associated with agro-food activity, such as soil and agricultural science, animal husbandry, horticulture, viticulture, farm management and agribusiness, food processing technology, agriculture environmental studies, and pest and weed control, have the smallest numbers of enrolments and accounted for only 16% of graduations in 2012. Some of these disciplines also experience a decline in completion rates (DEEWR, 2013). Students from rural/regional areas are underrepresented in tertiary education. In 2007, for example, they accounted for 18% of those receiving higher education compared to a 25% share of rural/regional group in total population (Bradley et al., 2008).

Figure 5.13. Percentage of students in tertiary education enrolled in agriculture programmes and the share of agricultural employment in Australia, 1998-2012

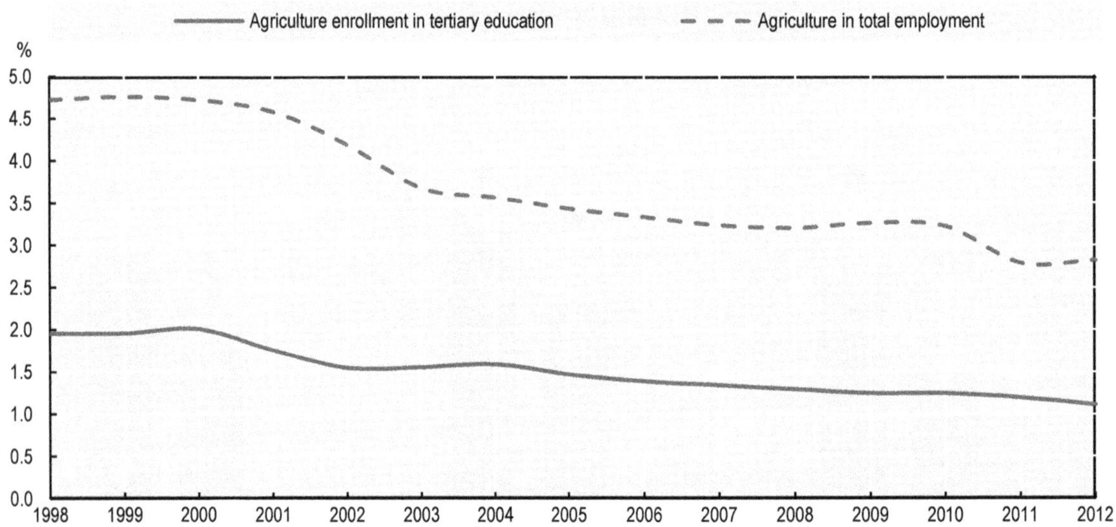

According to UNESCO's 2011 International Standard Classification of Education *(*ISCED*)*, agriculture as field of study includes the following disciplines: 62 Agriculture, Forestry and Fishery (agriculture, crop and livestock production, agronomy, animal husbandry, horticulture and gardening, forestry and forest product techniques, natural parks, wildlife, fisheries, fishery science and technology) and 64 Veterinary (veterinary medicine, veterinary assisting.

Source: UNESCO UIS Education and Literacy; OECD Stat.

StatLink http://dx.doi.org/10.1787/888933250034

The 2008 *Review of Australian Higher Education* stated that Australia was falling behind other countries in performance and investment in higher education, it also expressed a broad concern on whether Australia will have sufficiently qualified people to meet its medium and long-term needs (Bradley at al., 2008). This view referred to the evidence of persistent skill shortages in a number of higher education qualifications, and the "clear signs that the quality of the educational experience in Australia is declining". A modelling undertaken in the framework of this *Review* indicated that up to 2018 Australia's overall demand for people with qualifications would substantially exceed supply of students for higher education.[4] Furthermore, higher education faced an increasing difficulty in attracting and retaining high-quality academic staff, whereas the capacity of higher education to meet future labour market needs will be critically affected by the quality and capacity of the academic workforce.

Concerns about supply of skills are equally raised by the representatives of the agro-food industry, as expressed in their submissions to the Senate inquiry on skills to support future demand in agriculture and agribusiness (Agrifood Skills Australia, 2011; Agrifood Skills Australia, 2014; NFF, 2011; NSWG, 2012). The skill challenge is likely even more pronounced in these sectors, given the population outflow from rural areas, particularly among young people, and rapidly ageing workforce. The ageing also affects agricultural academic staff of which one half will be nearing retirement by 2018 (Agrifood Skills Australia, 2011). Demographic trends are compounded with less favourable performance in education of rural population compared to non-rural groups, such as performance of 15-year old rural students, representation in higher education, retention rates in education, and education attainments. This is occurring against a backdrop of a declining student demand for qualifications required by the agro-food sector, in particular for 'traditional' agricultural disciplines. The available estimates suggest that demand for university graduates for the agro-food sector is currently not met and the gap may be substantial, although any such estimates suffer from a lack of systematic analysis and adequate statistics.[5] The connection of skill supply and innovation is underscored by the innovation survey by the Australian Bureau of Statistics. According to it, 6% of surveyed businesses in the agriculture, forestry and fishery sectors with employment up to 4 people and almost 30% of businesses with employment between 5 and 19 people mention a lack of skilled persons among the principal barriers to innovation (ABS, 2014).

In general, Australia's flexible labour market and well-developed education sector provide essential mechanisms and capacity to respond to a mismatch in the supply and demand for skills in the agriculture sector. The skill challenge has been also tackled by policy. In the late 2000s, state and territory governments and Australian government agreed on nation-wide education targets. A number of underlying reforms and numerous initiatives were introduced at all levels of education aimed at enhancing skills across all of the economy, including in agro-food industries (Box 5.8). Overall, these reforms seek to achieve better coherence in the education system across jurisdictions, greater equity in education, and easier transition to further education or labour market, including through stronger integration of employers and educators. In 2014, the succeeding Australian government announced new reforms focussing on the vocational education and training system. The degree to which continuity of previous policy initiatives will be maintained through this process is not yet known (e.g. it was decided to discontinue the National Workforce Development Fund created in 2012 under the previous government to support employers in training new and up-skilling existing workers).

The agro-food industry has recently communicated its proposals for improving the provision of skills for the sector. Agrifood Skills Australia, one of the 11 Australia's Industry Skills Councils mandated to link the industry and educators, released the agro-food industry's vision of skill priorities and strategies for action. This is a broad framework spanning from business capacity to plan jobs and promote skills, through attracting new skilled generation, to enhancing and better utilising knowledge and skills of existing workforce. Beyond issues common to many sectors, this framework emphasises the need to tackle the specific challenge of boosting student preference for agro-food careers. It also stresses the importance of exploiting the potential of on-the-job training through informal education, given the important role of tacit knowledge in agriculture (Annex 5.A1).

Box 5.8. Australia's education targets and policies to enhance skill supply

The National Partnership on Youth Attainment and Transitions (2009) aims to retain youth in education, and to improve their transition to further education, training or employment to align to the Council of Australian Governments' goal of a 90% upper secondary (Year 12 equivalent) attainment rate by 2015.

In 2009, the Council of Australian Governments also set VET targets to be achieved by 2020, including: i) to increase to more than three-quarters the share of working age Australians with a Certificate III level qualification or higher (up from around half in 2009); and ii) to double the number of Diploma and Advanced Diploma completions.

In addition to the above VET targets, the Australian Government has also set two national targets for higher education: i) Attainment: by 2025, 40% of all 25-34 year olds will have a qualification at bachelor level or higher; ii) Participation: by 2020, 20% of higher education enrolments at undergraduate level will be people from low socio-economic status.

Australia was particularly active in multiple initiatives to improve the quality of vocational education and training (VET) to meet the demands of the labour market. A national VET regulator was established in 2011 to ensure adherence to national quality standards. The government also sought to improve apprenticeships through harmonisation of apprenticeship regulation across states. The National Partnership Agreement on Skills Reform (2008, renewed in 2012) outlines national reforms and includes a new entitlement to a subsidised training place for up to the first Certificate III qualification income-contingent loans for Diploma and Advanced Diploma qualifications. This agreement aims to support an estimated 375 000 additional students over the period of 2012-17 to complete their qualifications, with a focus on high level skills and disadvantaged students. The National Agreement for Skills and Workforce Development (2008) provided state/territory governments with greater flexibility to target funding to address individual state needs and to achieve agreed VET outcomes. The Australian government was to invest around AUD 15 billion over four years to assist all working-age Australians, including those employed in agriculture, fisheries and forestry, to access the skills and qualifications they need. This included AUD 7.2 billion to be made available to the states and territories for the vocational education and training (VET) sector and a further AUD 1.75 billion over four years to drive reform of the national training system.

In 2014, the new Australian Government initiated a new reform of the VET system. Several areas are identified: examining the standards for VET providers and regulators; reducing the burden on the VET sector arising from the constant updates to training packages; ensuring industry involvement in policy development and performance oversight, and streamlining governance arrangements and committees.

Source: OECD (2013b), *Education Policy Outlook: Australia*; Australian Government Department of Industry.

In conclusion, concerns persist about inadequate supply of skills in the agro-food industry, which may potentially limit the capacity of the sector to develop and uptake innovations. The degree and the character of demand-supply skill gap, as well as how it may evolve in the future, are unclear due to a lack of appropriate data and research. Demographic trends in rural areas, combined with less favourable performance in education of rural population compared to non-rural groups add complexity to the issue Australia's flexible labour market and education sector have the capacity to address the mismatch in the supply and demand for skills in the agriculture sector. Since the late 2000s, Australian and state and territory governments have been also active in addressing the skill challenge. The new Australian government has announced its education reform intentions and launched a number of reviews to support reform decisions. The assessment of the effectiveness of previous reforms is difficult due to short period of their implementation, while the degree to which the continuity of previous initiatives will be maintained is yet known. In any case, the evidence of the current and the probability of future skill shortages in the agri-food industry highlight the need for a more responsive system. Competition for scarce labour with more attractive and dynamic sectors is a challenge beyond the education system alone. The industry itself has a role to play to make agricultural and agri-food careers more attractive.

Notes

1. Proficiency is considered as a continuum of ability. Levels do not indicate minimum proficiency levels, but are used to help interpret results. Proficiency is described in terms of a scale of 500 points divided into levels. Each level summarises what a person with a particular score can do. For example, for problem solving in technology rich environment, adults at level 1 can only use widely available and familiar technology applications, such as e-mail software or a web browser, to solve problems involving few steps, simple reasoning and little or no navigation across applications; at level 2, adults can complete problems that involve a small number of computer applications, and require completing several steps and operations to reach a solution; while at level 3 can complete tasks involving multiple computer applications, a large number of steps, and the discovery and use of ad hoc commands in a novel environment (OECD, 2013c).

2. One exception is performance of Australia's rural students' in mathematics where they show the same score as the rural students on average across OECD area.

3. There are indications that the decline in the number of students choosing to study agriculture may have slowed. According to the Australian Department of Education, the higher education sector reported an 11% increase in preferences for agriculture-related courses in 2013 (Australian Government, 2014e).

4. This modelling was done by an advisory group Access Economics in 2008; the core underlying assumption was that up-skilling of the workforce over time will be a major contributor to labour market demand. If up-skilling does not occur, demand for higher qualifications may be more subdued (Bradley et al., 2008).

5. In the area of agronomy, it is estimated to be around ten jobs for every person with agricultural education and training (NSWG, 2011 citing Grain Producers Australia). According to the estimate of the Australian Council of Deans of Agriculture made in 2009, there may be as much as six-fold gap between demand and graduate numbers (Agrifood Skills Australia, 2011).

References

ABS (2012), Australian Social Trends, cat. No. 4102.0, Australian Bureau of Statistics, December 2012. http://www.abs.gov.au/AUSSTATS/abs@.nsf/Lookup/4102.0Main+Features10Dec+2012#AN%20AGEING

ABS (2013a), Measures of Australia Progress, 2013 – Data for Population, cat. No. 1370, Australian Bureau of Statistics, http://www.abs.gov.au/ausstats/abs@.nsf/Lookup/by%20Subject/1370.0~2013~Download%20Summary~Data%20and%20downloads~10

ABS (2013b), Measures of Australia Progress, 2013 – Progress Indicators for Jobs, cat. No. 1370, Australian Bureau of Statistics, http://www.abs.gov.au/ausstats/abs@.nsf/Lookup/by%20Subject/1370.0~2013~Download%20Summary~Data%20and%20downloads~10

ABS (2013c), Population Projections, Australia, 2012 (base) to 2101, cat. No. 3222.0 Australian Bureau of Statistics http://www.abs.gov.au/AUSSTATS/abs@.nsf/DetailsPage/3222.02012%20(base)%20to%202101?OpenDocument

ABS (2014), Innovation in Australia Business 2012/13, cat. No. 81580, Australian Bureau of Statistics, August. http://www.abs.gov.au/ausstats/abs@.nsf/mf/8158.0

Agrifood Skills Australia (2011), *Submission to the Senate Education, Employment and Workplace Relations Committee in relation to Its Inquiry into: Higher Education, and Skills Training to Support Future Demand in Agriculture and Agribusiness in Australia*, November 2011.

Agrifood Skills Australia (2014), 2014 *Environmental Scan of the Agrifood Industry*, Kingston ACT.

Australian Department of Employment (2014), *Labour Market Assistance Outcomes*, March 2014. https://employment.gov.au/labour-market-assistance-outcomes-reports.

Australian Government (2014a), *Australian Jobs 2014*, Australian Government, Department of Employment, Canberra.

Australian Government (2014b), Migration Blog – Agriculture and Migration, Department of Immigration and Border Protection, http://migrationblog.immi.gov.au/2011/07/29/agriculture-and-migration/

Australian Government (2014c), *Skill Shortages Australia 2013-2014*, Australian Government, Department of Employment, Canberra.

Australian Government (2014d), *State and Territory Migration Summary*, Australian Government, Department of Immigration and Border Protection, 31 March 2014, Canberra.

Australian Government (2014e), *Agricultural Competitiveness Green Paper*, Australian Government, Department of the Prime Minister and Cabinet, October 2014, Canberra.

Business Council of Australia (2014), "*Building Australia's Comparative Advantages*", Discussion Paper, July 2014, Melbourne.

Bradley, D., P. Noonan, H. Nugent, B. Scales (2008), Review of Australian Higher Education. Final Report, Commonwealth of Australia, December 2008, Canberra.

Collins, J., B. Krivopkapic-Skoko, D. Monani (2014), "New Immigrant Farmers in Australia: A contemporary Overview" Paper abstract presented at the 20th Bi-Annual Conference of the Asian Studies Association of Australia (ASAA),

DEEWR (2013), *Higher Education Collection*, Department of Education, Employment and Workforce Relations, Canberra.

Infrastructure Australia (2013), *National Infrastructure Plan June 2013*, Canberra, 2013 http://www.infrastructureaustralia.gov.au/coag.

Linehan, V, S. Thorpe, N. Andrews, Y. Kim, and F. Beaini (2012), *Food Demand to 2050: Opportunities for Australian Agriculture*, ABARES conference paper 12.4, Canberra, March 2012. http://data.daff.gov.au/data/warehouse/Outlook2012/fdi50d9abat001201203/Outlook2012FoodDemand2050.pdf.

NFF (2011), *Submission to the Senate Education, Employment and Workplace Relations Committee in relation to Its Inquiry into: Higher Education, and Skills Training to Support Future Demand in Agriculture and Agribusiness in Australia*, November 2011.

Nguyen, N., L. Hogan, K. Lawson, P. Gooday, R. green, K. Harris-Adams, and T. Mallawaarachchi (2013), *Infrastructure in Australia's Food Industry: Preliminary Economic Acessment*, ABARES Research Report 13.13, Canberra, November 2013 http://data.daff.gov.au/data/warehouse/9aap/2013/iafipead9aap_20131105/infaAustFoodInd_prelimEcoAssess_v1.0.0.pdf.

NSWG (2012), *Review into Agricultural Education and Training in New South Wales – Issues Paper*, New South Wales Government, Australia.

OECD (2010a), *Measuring Innovation: A New Perspective,* OECD Publishing, Paris. DOI: http://dx.doi.org/10.1787/9789264059474-en.

OECD (2010), *OECD Economic Surveys: Australia 2010*, OECD Publishing, Paris. DOI: http://dx.doi.org/10.1787/eco_surveys-aus-2010-en.

OECD (2012a), *Activating Jobseekers: How Australia Does It*, OECD Publishing, Paris. DOI: http://dx.doi.org/10.1787/9789264185920-en.

OECD (2012b), *OECD Economic Surveys: Australia 2012*, OECD Publishing, Paris. DOI: http://dx.doi.org/10.1787/eco_surveys-aus-2012-en .

OECD (2013a), *Education at a Glance 2013 - Country notes and key fact tables: Australia*, OECD Publishing, Paris. http://www.oecd.org/edu/educationataglance2013-countrynotesandkeyfacttables.htm.

OECD (2013b), *Education Policy Outlook: Australia*, OECD Publishing, Paris. http://www.oecd.org/edu/highlightsaustralia.htm

OECD (2013c), *OECD Survey of Adult* Skills (PIAAC) 2013, http://www.oecd.org/site/piaac/

OECD (2013d), OECD Programme for International Student Assessment (PISA), 2013, http://www.oecd.org/pisa/

OECD (2014a), *OECD Education at a Glance database*, 2014, http://www.oecd.org/education/eag.htm.

OECD (2014b), Online OECD *Employment Database*, http://www.oecd.org/employment/emp/onlineoecdemploymentdatabase.htm.

Productivity Commission (2014), *Public Infrastructure, Draft Inquiry Report*, Volume 1, March, Canberra.

Tan, Y., S. Richardson, L. Lester. T. Bai and L. Sun (2009), *Evaluation of Australia's Working Holiday Maker (WHM) Program*, National Institute of Labour Studies, Flinders University, Adelaide. February

WDI (2013), *World Development Indicators Database*. World Bank, http://data.worldbank.org/data-catalog/world-development-indicators.

World Bank (2013), *Doing Business 2014*, World Bank and the International Finance Corporation, World Bank World Economic Forum (2013), *The Global Competitiveness Report 2013-2014*: Full data Edition, Geneva 2013. http://reports.weforum.org/the-global-competitiveness-report-2013-2014/#=.

World Economic Forum (2013), *The Global Competitiveness Report 2013-2014*: Full data Edition, Geneva 2013. http://reports.weforum.org/the-global-competitiveness-report-2013-2014/#=.

Annex 5.A1

Agrifood industry workforce planning and skills priorities, key strategies for action and policy levers

Industry priorities	1. Building world-class business management capability	2. Attracting a new generation of motivated, skilled and smart workers	3. Building higher level knowledge and skills within the existing workforce	4. Increasing industry adoption rates of new technologies and research outcomes	5. Utilising the skills of existing workers and lifting retention rates
Key strategies for action by industry, governments and tertiary system	**1.1** Build employers' high-level business management capability, entrepreneurial and innovation capacity, export readiness and cultural competence through Skill Sets and undergrad programs developed by collaborative VET/ higher education/ service provider partnerships	**2.1** Develop common, national branding for the shared promotion of agrifood careers to convey the professional, contemporary and diverse face of industry, its job roles, career pathways and global opportunities	**3.1** Build higher level skills within the workforce through delivery of Skill Sets, units of competency and full qualifications	**4.1** Continuously improve nationally endorsed qualifications, Skill Sets and units of competency to reflect latest skills and knowledge	**5.1** Build sustainable, grass roots, 'skills formation strategies' and regionally based 'skills eco-systems'
	1.2 Build employers' workforce planning capabilities in effective job design, work organisation and the principles of 'employer of choice'	**2.2** Build attractive, broad based learning experiences into industry for school students through the Australian Curriculum, VET in schools and industry placements	**3.2** Remove duplication of effort and resources through harmonisation of key industry certification/licensing requirements with nationally endorsed qualifications	**4.2** Increase industry adoption rates of CRC/ RDC/ CSIRO outputs through collaboration and active partnerships with VET and universities	**5.2** Establish, brand and drive widespread formal recognition of existing workers' skills
	1.3 Build industry bodies' capability and role in promoting workforce planning and skills development to their members as a core business strategy	**2.3** Establish and pilot a contemporary, flexible model for agrifood technician/ paraprofessional cadetships in new and emerging job roles	**3.3** Build enterprise capability in key areas of industry reform through delivery of relevant units of competency/ Skill Sets	**4.3** Establish formal articulation arrangements and pathways between VET and higher education sectors	**5.3** Identify and promote skill based career pathways for existing workers within and across agrifood sectors

	1. Building world-class business management capability	2. Attracting a new generation of motivated, skilled and smart workers	3. Building higher level knowledge and skills within the existing workforce	4. Increasing industry adoption rates of new technologies and research outcomes	5. Utilising the skills of existing workers and lifting retention rates
		2.4 Restore universal incentives to Certificate II traineeships identified as key entry level occupations	**3.4** Increase language, literacy, numeracy and digital literacy capabilities of the existing workforce as a platform for further skills development		**5.4** Develop innovative responses to the skill needs of casual, contract and seasonal workers
		2.5 Establish, brand and widely promote a series of national recognition programs to attract and optimise existing skills of Indigenous Australians, migrants, refugees, and resource workers re-entering the broader workforce			
Key policy levers	1. **Establish industry/ government co-investment in existing workers training** as a national funding principle to formally recognise the dual clients of the national training system and provide a balanced policy setting to National Training Entitlement for individual students; 2. Establish national commitment to the **delivery of publicly funded Skill Sets and units of competency** to meet the identified needs of individual enterprises and learners; 3. Drive collaborative **partnerships between the publicly funded research/innovation system and VET system/ universities** to enable broader exposure of new knowledge and practices, and dramatically improve 'speed to market' of new practices.				

Source: Agrifood Skills Australia (2014), *2014 Environmental Scan of the Agrifood Industry*, http://www.agrifoodskills.net.au/?page=Escan14.

Chapter 6

Australian agricultural policy: Structural change, sustainability and innovation

This chapter provides an overview of domestic and trade agricultural measures, outlining those supporting investment, the adoption of innovation or environmental practices, and the development of bio-products. It also reports trends on the level and composition of support and discusses the likely impacts of agricultural policy measures on structural change, environmental performance and innovation in the sector.

The statistical data for Israel are supplied by and under the responsibility of the relevant Israeli authorities. The use of such data by the OECD is without prejudice to the status of the Golan Heights, East Jerusalem and Israeli settlements in the West Bank under the terms of international law.

Agricultural policy includes measures specifically designed for and applied to the agricultural sector. This distinguishes agricultural policy from general policies reviewed in the previous chapters which apply to agriculture and other sectors as parts of the overall economy. This section first discusses the objectives of Australia's agricultural policy and then looks at domestic price policy, payments based on various parameters of agricultural production or on other parameters inherent to agriculture, such as the environmental condition of land and water. Next, trade policy measures associated with domestic agricultural policies are examined. Finally, the extent to which Australia's agricultural policy is oriented to the support of agriculture's long-term productivity is evaluated based on OECD support indicators.

Agricultural policy objectives and instruments

As part of the broad economic policy reforms of the 1980s and 1990s, Australia implemented a phased dismantling of supply and price regulation arrangements in key commodity sectors (Annex 6.A1). Today Australia's agriculture sector is market oriented. The farming and agro-industry sectors operate overwhelmingly with no market price support, with domestic and international prices closely aligned. Agricultural policies have refocused on assisting farmers to manage various production risks, encourage increases in productivity and efficiency, and on the sustainable use of resources, in particular water.

The main support instruments applied by the federal and state governments across the range of programmes include income support payments, concessional loans and grants. Agricultural producers may also benefit from tax concessions aimed at smoothing income variations, facilitating structural adjustment and encouraging sustainable farming practices. Support is also provided in the form of (co)financing of R&D, extension and advisory services.

Australia is at the moment when its future agricultural policy is being defined as part of the broader policy agenda to boost productivity. A *Green Paper on Agricultural Competitiveness* was released in October 2014 summarising policy issues raised and policy proposals from over 700 public discussion submissions (Australian Government, 2014a). The policy issues presented in this document revolve around areas of infrastructure development; finance, business structures and taxation; competition and regulation; education and skills; resource management, research and development and other. This document is to feed into the *White Paper on the Competitiveness of the Agriculture Sector*, which will define government's future policy measures and related commitments.

Domestic agricultural policy

In terms of overall spending, the most significant directions taken by Australian agricultural policy represent support to the agriculture innovation system, tax concessions, infrastructure development, support to sustainable farming, and drought assistance (Figure 6.1). This section is focussed on the principal agricultural policy groups, while support to R&D and extension are reviewed in the section on agricultural innovation system.

Figure 6.1. Main groups of agricultural programmes in Australia, 2011-13 average

AUD million

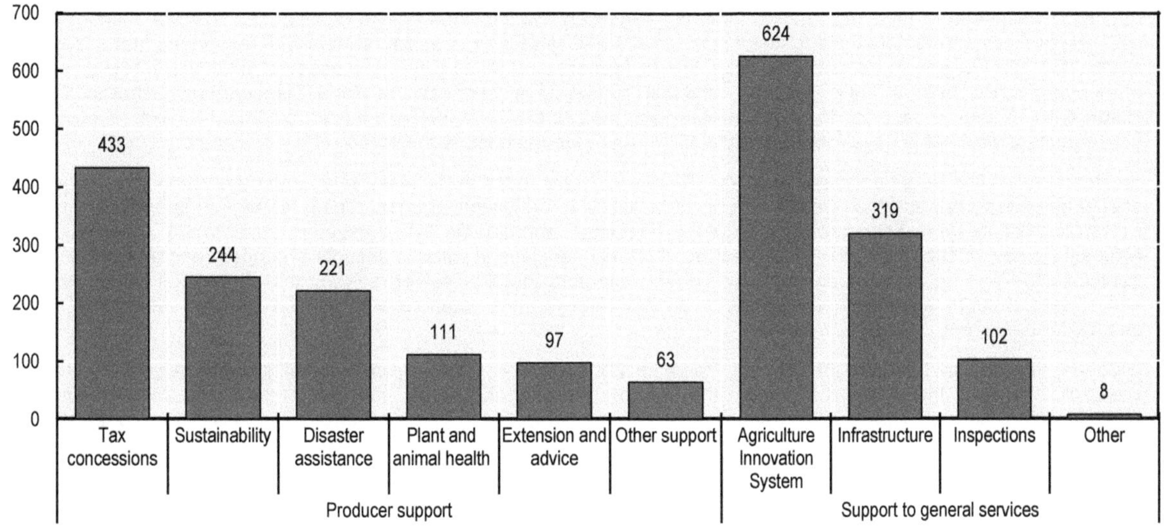

Source: OECD PSE/CSE database, 2014, www.oecd.org/tad/agricultural-policies/producerandconsumersupportestimatesdatabase.htm.

StatLink http://dx.doi.org/10.1787/888933250047

Disaster assistance[1]

Extreme variability of climate leads to the occurrence of severe weather events in Australia, such as cyclones, bushfire, landslides, hail storms, floods and droughts. Climatic risks in agriculture are addressed by various assistance programmes implemented by state and federal governments.

The Natural Disaster Relief and Recovery Arrangements (NDRRA) is a long-standing framework providing assistance with respect to a broad range of disasters, excluding drought, human or animal epidemic. The NDRRA provides partial reimbursement for damage arising as a direct result of a natural disaster. It does not provide compensation for complete losses and where insurance can cover the losses, assistance is generally not available. The NDRRA is a policy framework under which state and territory governments are given appropriate flexibility to develop their own programmes and measures under four main categories. Categories A and B are activated automatically by the state, with Category C and D assistance activated following Prime Ministerial agreement to exceptional circumstances. The federal government partially reimburses state/territory expenditures on an event if they exceed certain thresholds. The measures include assistance to individuals, families, communities, primary producers, small businesses and not-for-profit organisations, including one-off payments, assistance for clean-up and restoration costs, concessional rate interest loans for direct damage and loss of income, and freight subsidies.

The government's approach to drought was to define it as a natural disaster and provide automatic relief through subsidies on farm inputs. In 1992 the Commonwealth and state governments agreed to the implementation of a new National Drought Policy (NDP) that recognised the nature, effects and duration of drought were different compared to other natural disasters. The NDP was national in terms of its funding, design and decision making, the objectives of which were to: encourage primary producers and other sectors of rural Australia to adopt a self-reliant approach to manage climate variability; maintain and protect Australia's agricultural and environmental resource base during periods of extreme stress; and ensure early recovery of agricultural and rural industries that are consistent with long-term sustainable levels. Amid increasingly variable climate, the Australian government has continued to pursue reform of drought assistance to help farm households to focus on risk management and preparedness (Box 6.1).

Box 6.1. National Review of Drought Policy in Australia

The types of assistance originally provided under the National Drought Policy included: income support to cover day-to-day living expenses of farm households suffering extremely low incomes due to disaster (Exceptional Circumstances (EC) Relief Payment); grants for exit, re-training, advice and relocation to assist farmers with non-viable farms to exit the sector (EC Exit Package); interest rate subsidies to farmers and small rural businesses in financial difficulty resulting from drought (EC Interest Rate Subsidy); and tax concessions, such as the Farm Management Deposits Scheme (Table 6.A1.2) and measures to increase farmers' awareness and preparedness to drought risks. Since 1992, subsequent governments have maintained the NDP, but made changes to the programmes delivered under it.

Several reviews of drought policies during the 1990s and 2000s found that interest rates and transaction-based subsidies should be phased out and that greater emphasis be given to programmes encouraging farmer preparedness. By 2008, EC arrangements were acknowledged as being no longer appropriate. The Australian government commissioned a National Review of Drought Policy to assess the economic, social and climatic aspects of drought and drought policy. In particular, the findings in the economic assessment reiterated those of earlier reviews: that the NDP's EC declarations and related drought assistance programmes do not help farmers improve self-reliance, preparedness and climate change management.

Following the National Review of Drought Policy in 2008–09 and the two-year pilot of drought reform measures in Western Australia, the Australian, state and territory primary industries ministers signed the Intergovernmental Agreement on National Drought Program Reform (IGA) in May 2013. This Agreement formalised the commitment of Australian, state and territory governments to implement a new approach to drought programmes from 1 July 2014. The new approach replaces the previous arrangements and recognises that farm households can experience hardship for a range of reasons, not only drought. The IGA includes: a farm household support payment (Farm Household Allowance); continued access to the Farm Management Deposits Scheme and taxation measures; a national approach to farm business training; a co-ordinated, collaborative approach to providing social support services; and tools and technologies to inform farmer decision making.

The Intergovernmental Agreement continues to give government the flexibility to introduce additional support measures if needed in a drought. It contains an agreed set of principles and a process to guide governments in determining support in difficult times so that any drought support measures remain consistent with the intent of reform. In February 2014, the Australian government announced a drought assistance package worth AUD 320 million (USD 309 million) to support those farm businesses, families and rural communities facing hardship brought on by drought. Key components of the drought package include: more generous criteria for accessing income support for farmers; drought concessional loans to help eligible farm businesses cope with and recover from the effects of drought; assistance to drought-affected farm businesses for the installation of water-related infrastructure; assistance to reduce the impact of pest animals in drought-affected areas; and financial support for increased access to social and mental health services in communities affected by this drought.

Under the 2014/15 drought package, AUD 280 million (over three-quarters of the total budget) is allocated to drought concessional loans. This represents reduced-interest five-year loans that can be used: (i) for debt restructuring; (ii) to continue funding operating expenses; and (iii) for drought recovery and preparedness activities. The inclusion of concessional loans into the drought package seems to have been a situational decision amid considerable drought hardship. This approach, however, contradicts the argument against continuing business credit support developed during the drought policy review and which led to the decision to eliminate the EC Interest Rate Subsidy. The arguments against the continuation of business credit support pointed to its perverse incentives and unintended outcomes, especially in that recipients of credit assistance are given incentives to build up debt, may be less responsive to drought conditions, and have fewer incentives to diversify their activity to manage income risk. Finally, interest subsidies can become capitalised into asset values and penalise non-assisted farmers and new entrants (Productivity Commission, 2009a).

Sources: Australian Government (2014b), Intergovernmental Agreement on National Drought Reform; Australian Department of Agriculture, on-line information portal http://www.agriculture.gov.au/agriculture-food/drought/assistance/concessional-loans; Gray, E.M et al. (2014), *Australian Agricultural Productivity Growth: Past Reforms and Future Opportunities*, http://daff.gov.au/abares/publications.

In the 2000s, Australian farmers were exposed to significant yield variability which by far exceeded that in European countries; these risks were also systemic, i.e. farmers were affected in many different locations simultaneously. In response to increased income risks, farmers adjusted input use, crop rotations, and product mixes. In some cases, this has resulted in a more efficient mix of factors and inputs and thus, to productivity improvements. In others, the changed practices led to sub-optimal combinations of factors and inputs that slowed down productivity growth. The prolonged period of droughts probably had also an impact on farmers' long-term decisions making some reluctant about longer-term and new undertakings, such as investing in new methods and technologies. This could also explain the observed decrease in the rates of technical change and the slower catch-up of other farms which, in turn, lead to a slow-down of agricultural productivity growth rates.

Australia's agricultural risk management policy plays a particular role from the perspective of innovation due to its possible effects on farmer long-term decisions and the willingness to invest. Providing farmers with the information, knowledge and skills to prepare and adapt to risks helps to expand farmers' planning horizons and thus their willingness to invest and innovate. This in turn creates a virtuous circle when innovation becomes itself a means to manage production and market risks. In this sense, the pledge of the 2013 national drought programme reform to support self-reliant approaches to business risk management, as opposed to coping with business losses, is conducive to increased innovation. However, the first drought package for 2014/15 allocated the majority of funding for concessional credit to disaster-affected farmers to cope with the bank debts and cash flow deficits. This decision was taken under the conditions of considerable drought hardship. However, there is a long-run risk in departing from the original commitment of drought policy reform to focus on facilitating farmers' adaptation and to "avoid government being positioned as the "business lender of last resort" (Intergovernmental Agreement on National Drought Reform, 2014). Sustaining this commitment in the implementation of the new drought policy would go towards creating stronger incentives for farmers to undertake innovation and the sector to adjust. Beyond on-farm practices, measures to facilitate water trade and assistance to develop new risk management tools (e.g. index-based insurance) can enhance farmer's adaptation to climate risks.

Plant and animal disease

Plant and animal health is another risk-related area of Australian agricultural policies. While the states and territories implement their own disease and pest control programmes, joint Australian, state and territory programmes focus on particular national or regional issues, such as weeds of national significance, Australian plague locusts, or responses to emergency animal diseases and incursions of plant pests. Such programmes can include supporting research and capacity building required to manage these risks. Catastrophic damage caused by emergency animal diseases or plant pests is not considered a natural disaster or exceptional circumstance. Australia applies a "biosecurity partnership" approach to such risks through joint government-industry cost-sharing agreements that respond to emergency animal disease, plant pest and environmental incursions. These arrangements set out the roles, responsibilities and cost-sharing arrangements for their signatories, which include developing response plans and, where applicable, owner reimbursement costs. These emergency response arrangements are complementary to the off-shore, border, and domestic biosecurity measures falling under the sole responsibility of the Australian government (see biosecurity and quarantine regulation in Section 3.1.3), as well as to state and territory programmes.

Sustainability and climate change

Agricultural policy over two recent decades has been oriented to reducing the considerable environmental footprint of agriculture and to climate change issues. As the driest inhabitable continent, Australia is vulnerable to the potential effects of climate change. Some agricultural practices (e.g. overgrazing, land clearing, and inefficient irrigation) have cumulated negative effects, such as aggravated soil salinity and acidity, erosion and pest damage. Agro-environmental measures are implemented as part of several major nation-wide frameworks, including the National Water Initiative, the National Land Care Programme, Carbon Farming Futures, and the Carbon Farming Initiative (the latter currently being merged with the new nation-wide programme the Emissions Reduction Fund, see Box 6.2). These frameworks use various channels to support activities by farmers, industries, regions and communities related to environmental improvements, mitigation of, and adaptation to climate change. Funding is provided for regional, community and farm-level projects. The assistance also includes regulatory support for the operation of water and carbon emissions markets, training and advice, provision of technical information, institutional support to environmental bodies, and research into new technologies and improvements of practices (Table 6.A1.1). Several recently expired programmes or that have been merged into broader current initiatives addressed specific adjustment pressures related to climate change; for example, the exit grant package for producers in the Murray-Darling Basin affected by drought and climate change.

Box 6.2 Recent developments in Australia's carbon emissions policy

Australia is the sixth top emitter of total greenhouse gases (GHG) among OECD countries and has the second highest level of emissions per unit of GDP (OECD, 2013b). Agriculture is a source of emissions, but it also provides a carbon sink function under certain management practices. Agriculture contributes 15% to total GHG emissions, one of the highest shares across the OECD countries. Almost three-quarters of all agricultural GHG emissions come as methane from ruminant livestock and manure management, while agricultural soils, particularly due to application of nitrogenous fertilisers, are the principal source of nitrous oxide. In the 1990s and 2000s, there has been progress in reducing GHG emissions from agriculture due to economy-wide measures, such as the adoption of national air quality standards setting ambient concentration limits for a number of conventional pollutants, the halting of land clearing, as well as the encouragement of emission-reducing farming practices. Important policy developments were also taking place recently (OECD, 2013c).

In July 2012, Australia introduced a Clean Energy Plan, including a carbon pricing mechanism ("carbon tax"). This mechanism required entities which emitted over 25 000 tonnes of carbon dioxide equivalent greenhouse gases per year to surrender emissions permits. The transport and agriculture sectors were exempt from this scheme. A move to an emissions trading scheme was intended as of July 2015 whereby emitters would have been enabled to trade their allowable (capped) emissions. Thus, firms that needed to increase their volume of emissions would have been obliged to buy permits from those who required fewer permits.

However, GHG emissions policy became subject to significant change under the new government. Following its election commitment, the carbon pricing mechanism was repealed in July 2014. A suite of new measures was proposed called the Direct Action Plan with the Emissions Reduction Fund (ERF) at its centre. Incentives will now be provided through the operation of the ERF across the economy. This fund will use the framework already applied in agriculture – the Carbon Farming Initiative. The companies with emissions abatement projects which submit these to the Clean Energy Regulator become eligible for the funding from the ERF. The funding will be allocated to projects capable of achieving maximum abatement at least cost. Project eligibility will be assessed, and the projects will be ultimately accredited. Payments from the ERF to project proponents will be made upon delivery of abatement. The Carbon Farming Initiative ceases as a separate framework and will be merged into the Emissions Reduction Fund.

The ERF is a mechanism for crediting emissions reductions, whereby authorities issue credits for emissions reductions that are measured and verified by approved methods. Authorities will administer sealed bid auctions and purchase emissions reductions, on behalf of the government, at the lowest cost across the economy. The government will then enter into contracts with successful bidders. The contracts will provide a guarantee to businesses about the quantity and the price of carbon credits that the government will purchase on delivery. It is also intended to introduce a safeguard mechanism to ensure that the emissions reductions paid for through the Emissions Reduction Fund are not offset by increases elsewhere in the economy. The Australian Department of the Environment, considers that such a scheme can in principle have the same incentive effects as a carbon tax at the margin. Legislation implementing the ERF passed the Australian House of Representatives in June 2014 and is now before the Senate.1 Currently, consultations with business are underway on the development of the safeguard mechanism with a view to introduce it on 1 July 2015.

1. The Australian government introduced to Parliament legislation for the Emissions Reduction Fund (ERF) in the form of the Carbon Credits (Carbon Farming Initiative) Amendment Bill 2014. The amendments under this Bill concern: requirements for participation in the ERF and the expansion of eligible offsets projects; the process by which existing carbon farming projects and methods will transition into the ERF; establishment of the Emissions Reduction Fund Register; streamlining of assessment and estimation methods; auditing and reporting arrangements; process for selecting emission reduction projects; and other.

Source: Australian Department of Environment; Australian Department of Agriculture.

Tax concessions

Australia's tax regime incorporates various concessions to agricultural producers, which in a large part complement the policies described above. Tax concessions are available to farmers as generally provided to all businesses, or only to primary producers, SMEs, or agriculture. Some of the tax concessions are aimed at smoothing farm income fluctuations, thus operating as policy instruments to support producer income risk management. These include income tax averaging and the Farm Management Deposits Scheme. Other tax preferences are in place to facilitate structural adjustment. On the one hand, this includes exemptions from capital gains tax for exiting small businesses, also available to farmers; on the other hand, there are tax incentives for investments in specific activities. Another group of preferences is linked to environmental improvements, soil conservation, sustainable water use, and the establishment of carbon sink forests (Table 6.A1.2).

Farmers, along with other sectors and activities, also benefit from fuel tax credit through a reduced excise tax for several types of fuels used in off-road activities. Excise tax and Goods and Services Tax concessions are also granted for specific groups of food products.

Agricultural policy thus uses a diverse scope of tax concessions that affect incentives for technological change and the process of structural adjustment.

Other support

Several types of assistance are offered to producers in financial difficulty. *Farm Finance Initiative* supports farmers facing high levels of debt but who nevertheless demonstrate long-term viability. Eligible farmers are able to access short-term (up to five years) concessional loans. Rural Financial Counselling Service provides free financial advice for primary producers, fishers and small rural businesses experiencing financial hardship. From 1 July 2014, the Farm Household Allowance (FHA) is providing eligible farmers and their partners who are experiencing financial hardship with assistance and support to improve their long-term financial situation. Prior to the FHA, the Transitional Farm Family Payment and then the Interim FHA provided support to low income farm families to assist with basic household needs and to improve long-term financial security.

Agricultural trade policy

Australia's agricultural tariff protection is very low, with an average applied agricultural tariff for the sector (excluding forestry) of 1.2% in 2012 (WTO, 2011). Most products enter duty free, but some sensitive items, such as cheese (which is subject to a specific duty), and certain vegetables (mushrooms), nuts, fruit, oils and fats (subject to a 5% rate), receive some tariff protection. Australia maintains tariff-rate quotas for certain types of cheese. Imports are subject to technical regulations and biosecurity measures, involving the possibility of import prohibitions and restrictions (see biosecurity and quarantine regulation in Section 3.1.3).

Australia applies relatively low tariffs for capital and intermediate goods, but which are above the levels observed in Canada and with respect to capital goods, above the tariff levels in Japan, the United States and the European Union. Protection in Australia is somewhat skewed towards industrial goods as compared to agricultural ones (Figure 6.2).

Figure 6.2. Import tariffs for industrial and agricultural goods,[1] 2012

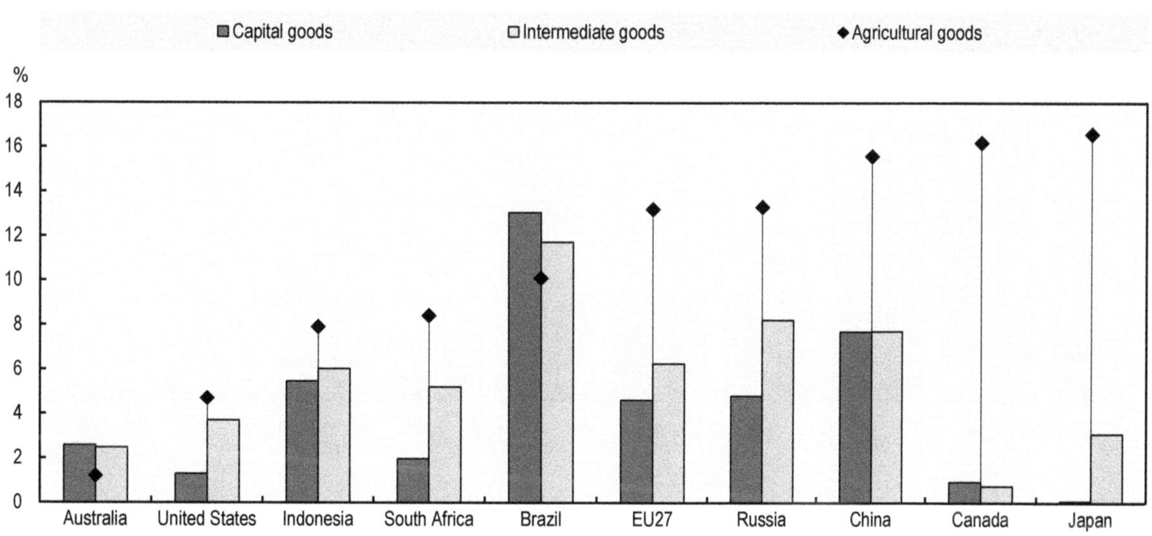

1. Simple average MFN tariffs, specific duties in *ad valorem* equivalents included. Tariff rates for non-agricultural products do not include specific duties.

Source: UNCTAD Trade Analysis Information System (TRAINS) and World Tarff Profiles, 2013.

StatLink http://dx.doi.org/10.1787/888933250053

Level and composition of agricultural support from the perspective of innovation

Domestic agricultural and associated trade measures affect farm investments and practices through a variety of instruments, with different impacts on innovation and sustainability. Several dimensions of agricultural policies are important in this context: (i) the extent to which market-distorting instruments are used to support producers; (ii) the extent to which they provide support to general services for the sector as opposed to support to individual producers; and (iii) the extent to which support is targeted to specific innovation and sustainability activity.

Support to agricultural producers: use of most distorting support

A key characteristic of agricultural policy from the perspective of innovation is the extent to which producer support relies on measures that distort agricultural output and input markets. OECD analysis on production, trade and income effects of agricultural support has shown that border protection, supply controls, output-based payments and variable input subsidies are potentially the most distorting. As such, these policies strongly diminish producer incentives to employ production factors more efficiently and to innovate so as to become more competitive. Distorting policies dampen incentives for innovation not only because they shield producers from competition, but also because they perform weakly in terms of generating income for long-term investment. It has been shown that most of the support provided through price interventions and variable input subsidies is captured in land rents by input suppliers and is lost as deadweight (Martini, 2011). Most distorting measures are not likely to be advantageous for investment because they encourage riskier behaviour by producers, making them more exposed to market and natural risks (OECD, 2011). On the other hand, broad-based income support decoupled from commodity production is more effective in transferring income to producers and thus increasing their capacity to invest and innovate. It also leaves more flexibility to producers to undertake new activities and switch to new products. If conditional on the adoption of environmentally-friendly practices, this support may help more sustainable resource use. However, even if decoupled from production choices, income support still slows structural adjustment needed to facilitate economies of scale, attract new entrants, and thus foster innovation and productivity growth.

Figure 6.3. Composition of Producer Support Estimate (PSE) by country, 2011-13

As a percentage of gross farm receipts

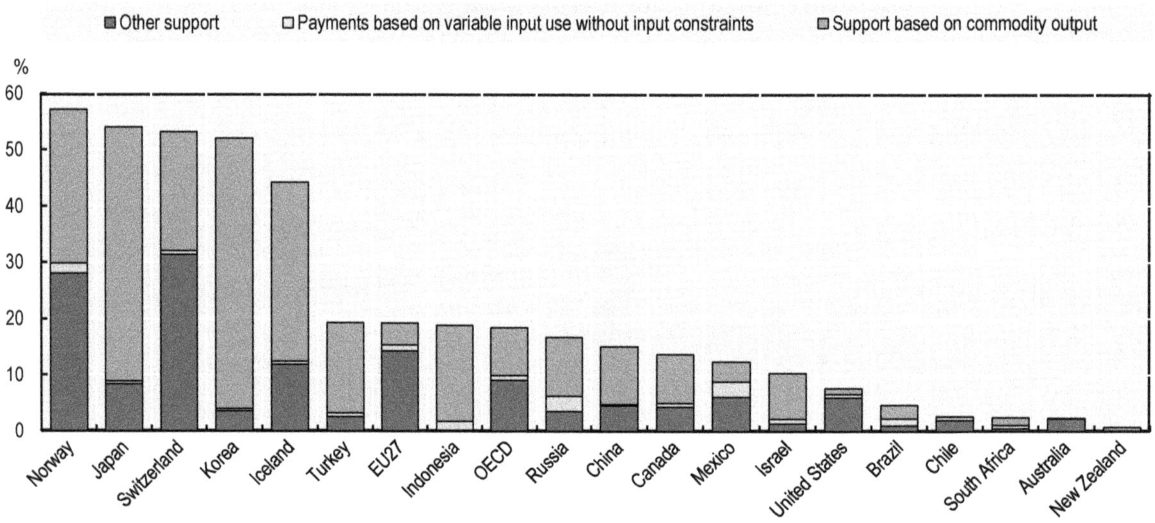

Source: OECD PSE/CSE database, 2014.

StatLink http://dx.doi.org/10.1787/888933250061

OECD indicators of agricultural support show that producers in Australia operate in a competitive environment with policies creating few distortions to producer incentives. The level of support to producers in Australia is one of the lowest across the countries covered by the OECD agricultural policy monitoring, with the Producer Support Estimate (PSE) at slightly above 2% of gross farm receipts in 2011-13 (Figure 6.3). Producer prices are closely aligned with international levels, as evidenced by the Nominal Protection Coefficients (NPC) equalling unity for all farm commodities included in the estimates. The low-distortive nature of agricultural policies in Australia also comes from the fact that potentially most distorting support accounted for only 3% of total support transfers to producers in 2011-13. This support was due to interest rate subsidies provided during this period as part of drought assistance.

Total support to the agricultural sector: Focus on innovation and productivity enhancement

Another key feature of agricultural policy from the perspective of innovation is the extent to which it is oriented towards supporting systems ensuring long-term productivity improvements. Such policy orientation can be revealed by the importance in the overall support of the investments in development of knowledge systems, infrastructure and institutions. These investments differ from assistance to individual producers (measured by the PSE) in that they support systems essential for the efficient functioning of the entire agro-food system and which provide with long-term benefits.

Total support to agriculture in Australia is relatively evenly distributed between transfers to individual producers and support to general services (Figure 6.4). As measured by the General Services Support Estimate, (GSSE), the latter accounted for 47% of total support to agriculture in 2011-13, while support directed to producers individually for 53%.

The GSSE indicator includes the financing of knowledge-related activities, such as agricultural research and education, as well as investments in infrastructure, land reform, inspection and control systems, activities on promotion and marketing of agricultural products, and the costs of maintenance of public stockholding systems. Within this broad range, activities supporting knowledge systems, infrastructure development and inspection and control systems contribute directly to fostering innovation and productivity in the agricultural sector. This funding, both at the Commonwealth and state and territory levels, is overwhelmingly directed for support to agro-food R&D and education activities, infrastructure development (largely, water-related), as well as biosecurity (Figure 6.4). Since the period 1999-2001, expenditure on hydrological infrastructure has tripled, reflecting efforts to modernise irrigation infrastructure in order to improve water use efficiency. As a result, the share of these expenditure in the GSSE has increased from 14% in 1999-2001 to 24% in 2011-13, while the share of expenditure on R&D decreased from 71% to 57% over the period (OECD, 2014c).

Australia is one of the few monitored countries with a relatively high share of total support directed to general services for agriculture. Only in New Zealand and Chile spending on general services makes up a higher share of total support to agriculture (over one half), with this percentage reaching no more than a quarter, and more often much less, in all other countries (Figure 6.5.A). The degree to which countries prioritise innovation and productivity-enhancing areas within their support to general services financing, however, varies widely. From this standpoint, Australia's support is overwhelmingly directed to investments in productivity-enhancing areas (Figure 6.5, Panel B).

Figure 6.4. Composition of Total Support Estimate (TSE) in Australia, 2011-13 average

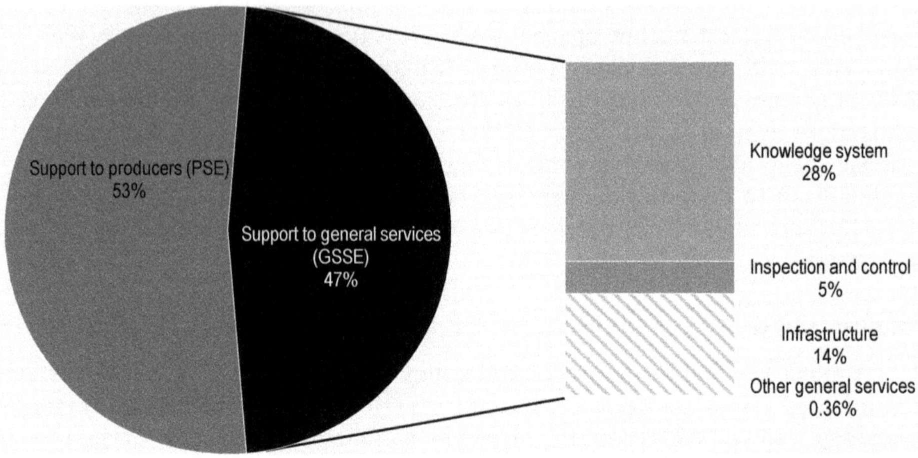

Innovation and productivity enhancing GSSE includes financing of knowledge and innovation systems, infrastructure and inspection and control systems. Other GSSE includes financing of marketing and promotion, public stockholding and miscellaneous expenditures.

Source: OECD PSE/CSE database, 2014.

StatLink http://dx.doi.org/10.1787/888933250074

Figure 6.5. Total support to agriculture (TSE) and general services (GSSE) in international comparison, 2011-13 average

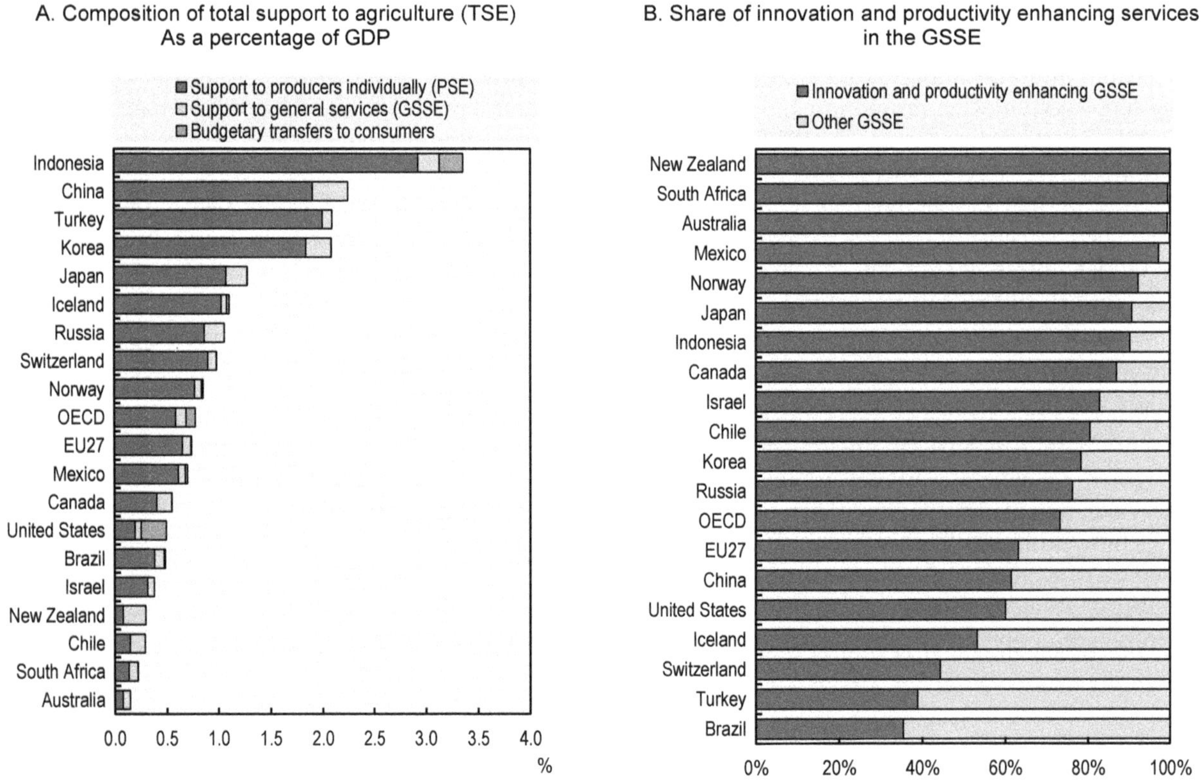

Source: OECD PSE/CSE database, 2014.

StatLink http://dx.doi.org/10.1787/888933250082

Note

1. A comprehensive review of Australia's agricultural risk management policies is contained in Kimura and Ánton (2011), and a comparative overview and assessment of risk policies can be found in OECD (2011).

References

Australian Government (2014a), *Agricultural Competitiveness Green Paper*, Canberra, October, https://agriculturalcompetitiveness.dpmc.gov.au/key-documents.

Australian Government (2014b), Intergovernmental agreement on national drought reform 2014, http://www.agriculture.gov.au/ag-farm-food/drought/drought-policy/drought-program-reform/iga-national-drought-program-reform.

Australian Treasury (2014), *The 2013 Tax Expenditure Statement*, the Treasury, Commonwealth of Australia, January 2014, Canberra.

Gray, EM, M. Oss-Emer, and Y. Sheng (2014), "Australian agricultural productivity growth: past reforms and future opportunities", ABARES research report 14.2, Canberra, February. Available at: http://www.daff.gov.au/abares/publications.

Kimura, S. and J. Ánton (2011), "Risk Management in Agriculture in Australia" *OECD Food, Agriculture and Fisheries Working papers,* No. 39, OECD Publishing, Paris. DOI: http://10.1787/5kgj0d8bj3d1-en

OECD (2011), *Managing Risk in Agriculture: Policy Assessment and Design*, OECD Publishing, Paris. DOI: http://dx.doi.org/10.1787/9789264116146-en.

OECD (2013b), *Environment at a Glance 2013: OECD Indicators*, OECD Publishing, Paris. DOI: http://dx.doi.org/10.1787/9789264185715-en.

OECD (2013c), *OECD Compendium of Agri-Environmental Indicators*, OECD Publishing, Paris. DOI: http://dx.doi.org/10.1787/9789264181151-en .

OECD (2014a), *Agricultural Policy Monitoring and Evaluation 2013: OECD Countries*, OECD Publishing, Paris. DOI: http://dx.doi.org/10.1787/agr_pol-2014-en .

OECD (2014c), *OECD PSE/CSE database*, www.oecd.org/agriculture/PSE

WTO (2011), *Trade Policy Review: Australia—Report by the Secretariat (Revision)*, Trade Policy Review Body, WT/TPR/S/244/Rev.1, 18 May, World Trade Organization.

Annex 6.A1

Australia's agricultural policy reform

Over the last 30 years the Australian economy was gradually transformed by a series of reforms that gained momentum from the early 1980s. By exposing Australian industries to greater international and domestic competition, and ensuring prices reflected actual costs, the reforms aimed to increase national living standards by increasing flexibility and productivity, promoting structural change and improving the competitiveness of firms and industries (Industry Commission, 1998). Australia's reforms took place at a time when the global economy was becoming more integrated, and as other countries were also initiating economic reforms.

Australian agriculture was lightly assisted compared with other sectors, in particular manufacturing, and also compared with agriculture in North America and Europe. Nevertheless, a range of measures were in place to maintain and stabilise farmer returns and to compensate for the costs of assistance provided to other sectors. These included marketing and price support schemes in sensitive industries such as dairy, sugar and tobacco, and subsidies to reduce farm input costs. Coupled with significant differences in rates of assistance across the sector, many industries faced distorted price signals that impeded industry adjustment and efficient resource use.

Historically, Australian government interventions in agriculture were largely concerned with improving farmers' welfare and increasing production and exports to earn foreign exchange. Australia's policy of developing the manufacturing industry through import replacement further increased this reliance, by encouraging local industries to focus on the domestic market. Other interventions aimed to stabilise and maintain farmers' incomes, assistance was also provided as compensation for the effects on farm input costs of industrial protection, most importantly, through domestic marketing and price support schemes and export price underwriting, import tariffs, income tax concessions, assistance to inputs and under-priced infrastructure services (e.g. irrigation services) and drought assistance.

Increasingly, the sectoral and economy-wide costs of assistance to agriculture caused stakeholders to question the effectiveness and efficiency of many agricultural policies. In the first instance, most assistance measures were not an effective means to improve producer welfare. Domestic pricing arrangements also reduced domestic consumption and welfare and contributed to raising costs levels generally. The structure of assistance was distorting resource allocation across the sector. Finally, agricultural policies and assistance measures were distorting farmers' incentives to find better ways of managing risks and to improve productivity more generally.

Reform of Australia's agricultural policies began in the early 1970s, when governments sought to limit the amount of budgetary assistance. Subsequent reforms aimed to make decision-making more responsive to market forces, and progressively reduced the level and narrowed the differences in rates of assistance across the sector. Agriculture was part of the economy-wide reforms of the 1980s. By 1995, all such assistance was removed for principal agricultural commodities. Tariffs were progressively phased down or out in other industries (dairy, dried vine fruits, sugar and wine). The fertiliser consumption subsidy was also removed in 1988 (Industry Commission, 1998). During the 1990s and 2000s, competitive conduct rules of the *Trade Practices Act 1974* (now the *Competition and Consumers Act 2010*) was extended to Statutory Marketing Authorities (SMA), resulting in the dismantling of the majority of SMAs. Other reforms from the 1990s onwards also encouraged greater market responsiveness, risk management and self-reliance. In particular, the 1990s saw a major shift in focus for drought policy in Australia.

Box 6.A1.1. Reforms to agricultural marketing arrangements

Competition has been gradually introduced into most agriculture industries where compulsory agricultural marketing arrangements previously had governed processes between the farm and (either or both) domestic and export markets. Key reforms include:

1970s	Wheat	Move from guaranteed to equalization prices; provision for 'grower to buyer' sales outside the pooling arrangements; home consumption price limited to wheat for human consumption and determined by a formula to take account of export prices
1980s	Dried vine fruits	End of price equalization arrangements in 1980
	Citrus	Decade-long phase down of tariffs from 30% to 5%, beginning in 1986; state marketing boards amalgamated, reducing geographical barriers to competition
	Cotton	Queensland Cotton Board deregulated in 1989
	Eggs	State-based production and pricing controls progressively withdrawn from 1989
	Sugar	Domestic administered price arrangements and export controls terminated by the Commonwealth in the late 1980s
	Wheat	Domestic market deregulated in 1989; grower levy fund introduced to replace the Commonwealth guarantee of Australian Wheat Board borrowing
1990s	Barley	Competition gradually introduced into domestic feed and malting barley marketing in South Australia and Victoria from 1998
	Dairy	Phased reductions in market support payments on export of dairy products
	Dried vine fruits	Commonwealth price equalization levy and statutory equalization of domestic sales removed in the early 1990s, as was the industry's exemption from section 45 of the Trade Practices Act (which reduced the scope for collusive price discrimination)
	Horticulture	Underwriting scheme for apples and pears terminated in 1990
	Tobacco	Local Leaf Content Scheme and the Tobacco Industry Stabilisation plan ceased in 1995; withdrawal of vesting powers in 1995
	Sugar	Import tariffs and domestic price supports removed in mid-1997
	Wheat	Australian Wheat Board converted from statutory authority to a grower-owned company in 1999
	Wool	Reserve Price Scheme ceased in 1991
2000s	Dairy	State-based controls over sourcing and pricing of market milk ceased in 2000; 9-year Dairy Industry Adjustment Package (DIAP) concluded in 2009
	Barley	South Australian single-desk arrangements terminated in 2007; Western Australian market deregulated in 2009 (allowing any number of licensed entities to export barley)
	Canola	Exports of canola and lupins deregulated in Western Australia in 2009 (traders no longer required to apply for licenses to export)
	Sugar	Queensland Sugar Limited lost its compulsory acquisition powers in 2006 and lost exemption from the Trade Practices Act in 2009
	Wheat	Bulk exports deregulated in 2008, meaning proposals to export bulk wheat no longer needed approval from the single desk seller (Australian Wheat Board)
Ongoing	Rice	New South Wales Rice Marketing Board still retains powers to vest, process and market all rice produced in New South Wales (around 99% of Australian rice is produced in this state)
	Potatoes	Western Australian Potato Marketing Corporation still controls the supply of fresh table potatoes in that state

Source: Gray et al. (2014) based on Industry Commission, Productivity Commission and WTO.

Table 6.A1.1. Main Australian environmental programmes and initiatives covering agriculture

Programme/activity	Environmental objectives	Content	Policy instruments	Start year
Carbon Farming Initiative (CFI), transitioning into the Emissions Reduction Fund (ERF)	• Reduction of GHG emissions	A voluntary offset scheme to facilitate the sale of carbon credits by reducing GHG emissions and storing carbon through changes in to agricultural and land management practices. Eligible on-farm activities that generate carbon credits are funded. When implemented, the Emissions Reduction Fund will apply to activities across the whole economy, and increase the range of eligible land sector activities. It will enable the government to purchase the carbon credits generated to meet Australia's abatement target (see Box 5.2 for more detail).	• Tradable rights/permits • Payments based on practices and delivery of abatement	2011
Carbon Farming Futures	• Reduction of GHG emissions	Funds research into new technologies and practices to emissions and store carbon in the landscape; assisting industry and farming groups to test and apply research outcomes; providing technical information and support to assist farmers, land managers and their key influencers to participate in land sector emissions management activities and the Carbon Farming Initiative (above).	• R&D, technical assistance • Tax concessions	2012
National Landcare Programme	• Adoption of sustainable practices • Land conservation • Conservation of biological diversity • Water quality	It supports the voluntary network of more than 6 000 groups across Australia and provides funding to support activities at the national, regional and local level. This funding gives farmers, landholders and community groups the ability to undertake on-ground action, which contributes to sustainable agricultural and environmental outcomes. The Programme also funds advice, information sharing, partnerships and networks across landcare and natural resource management stakeholders. Provides funding projects that target specific areas of sustainable farming and better resource management, precision farming and reduced tillage systems. Also funds farm projects on land conservation, biodiversity conservation and the stewardship of our natural environment. Provides funding for regional and community-based organisations to implement environmental projects, some agriculture-related projects funded concern, for example, soil erosion, wetland management, and carbon farming.	• Payments based on practices • Community based measures • R&D, technical assistance	1990s
National Water Initiative (NWI)	• Sustainable water extractions • Water quality • Addressing future adjustment issues	The NWI comprises eight key elements: i) water access entitlements and planning framework; ii) water markets and trading; iii) best practice water pricing; iv) integrated management of water for environmental and other public benefit outcomes; v) water resource accounting; vi) urban water reform; vii) knowledge and capacity building; and viii) community partnerships and adjustment.	• Water resource plans • Tradable water access rights (entitlements) • User-pays and cost recovery where possible • Payments based on practices • Community based measures • Technical assistance	2004

Source: Australian Department of the Environment; Australian Department of Agriculture; Gray et al. (2014); OECD (2014a).

Table 6.A1.2. Main tax preferences for primary producers[1] and the food sector in Australia

Concession/beneficiary	Content	Tax type
Primary producers		
Income tax averaging	Primary producers can elect to pay tax at tax rate based on their average income earned over the previous five income years. On balance, the saving from paying less tax in high income years outweighs additional tax paid in low income years.	Income tax
Farm Management Deposits Scheme (FMD)	The FMD scheme allows primary producers (with a limited amount of non-primary production income) to defer their income tax liability for primary production income earned. Deductions for the FMD may be claimed from the taxable income in the year of deposit, with subsequent withdrawals being subject to assessment in the year of withdrawal.	Income tax
Insurance recoveries	Insurance recoveries may be received in relation to livestock lost due to disasters or timber lost because of fire. Primary producers who receive such insurance recoveries can elect to spread the income equally over five income years, resulting in tax deferral.	Income tax
Income from double wool clips	As a consequence of drought, fire or flood, primary producers carrying on a sheep grazing business in Australia may conduct advanced shearing. In these circumstances, a woolgrower may elect to have the assessment of the profit from advanced shearing deferred to the succeeding income year.	Income tax
Income from forced disposal or death of livestock	Primary producers may experience forced disposal or death of livestock due to certain events, e.g. destruction of pasture due to drought, destruction of livestock for disease control, etc. Those who receive income from such disposals or deaths can elect to defer this income and use it to reduce the cost of replacement livestock. Alternatively, they can elect to spread profits between the income year of the disposal or death and the next four income years, or ten years in specific cases.	Income tax
Valuation of livestock from natural increase	Animals acquired by natural increase (i.e. new-born animals) may be valued at cost, in which case the taxpayer can use actual cost or costs prescribed by the regulations; the latter may be lower than the actual cost of production, giving a concessional tax treatment.	Income tax
Tax exemption for farm re-establishment grants	Re-establishment grants are provided, under prescribed industry re-establishment or exit grants (for example, the dairy, sugar and tobacco industry exit programs) o eligible farmers who choose to sell their farm and exit farming for at least five years. Such grants are exempt from capital gains tax.	Income tax
Accelerated depreciation schemes	Investments related to primary production may be eligible to certain general accelerated depreciation provisions and also specific to agriculture, such as the establishment of water facilities, establishment of horticultural plantations and vineyards, and horse breeding.	Income tax
Concessions for small businesses	Primary producers are eligible for various general income tax concessions for small business, including simplified depreciation schemes, deductions for capital expenditures, e.g. for electricity connections and telephone lines, tax breaks, etc.. These tax concessions apply to small business in general, while deductions for capital expenditure on telephone lines are specific lines are specific to primary producers	Income tax
Carbon sink forests	The cost of establishing trees in carbon sink forests was immediately deductible between 2007/08 and 2011/12 income years inclusive. After this initial period, such costs are deductible over 14 years and 105 days at a rate of 7% per annum	Income tax
Landcare deduction	Primary producers and users of rural land can claim a deduction for capital expenditure on a landcare operation. These operations may include soil conservation, prevention of land degradation or other related measures.	Income tax

Table 6.A1.2. Main tax concessions for primary agriculture and the food sector in Australia *(cont.)*

Concession/beneficiary	Content	Tax type
Sustainable Rural Water Use and Infrastructure Program	Taxpayers may choose to make payments received under the eligible agreements of the Sustainable Rural Water Use and Infrastructure Program free of income tax (including capital gains tax), with expenditures funded by such payments not being deductible.	Income tax
Carbon Pricing Mechanism (CPM) – agriculture	Emissions from agricultural activities were excluded from the CPM. The CPM was repealed in 2014.	CPM
Carbon Pricing Mechanism (CPM) – deforestation	Emissions from deforestation were excluded from the CPM. The CPM was repealed in 2014.	CPM
Fuel tax credits	Specified fuels used in off-road business activity are eligible for fuel tax credit (fuel tax credits are available to most businesses, not just primary producers).	Excise tax or customs duty
Goods and Services Tax (GST) on supplies of farm land	Certain supplies of farm land are GST-free. This includes: farm land supplied for farming on which a farming business has been carried on for at least five years and upon which a farming business is intended to continue to be carried on; and subdivided farm land that is potential residential land that is supplied to associates.	GST
Food producers		
Goods and Services Tax (GST) on food	Most food items for human consumption that are prepared and/or consumed at home are GST-free. Examples of GST-free food include fresh fruit and vegetables, fish, dairy products, bread and meat. Examples of GST-free beverages include unflavoured milk products, tea, coffee, water and fruit juices. In addition, generally the packaging used in the supply of GST-free food will itself be GST-free.	GST
Excise tax concessions	Excise tax concessions are applied on certain beers depending on concentration, container size and purpose, as well as brandy..	Excise tax

1. The definition of "primary producers" covers agriculture, forestry and fishing.

Source: Australian Treasury (2014); Australian Taxation Office.

Chapter 7

The Australian Agricultural Innovation System

This chapter outlines the role of a well-functioning agricultural innovation system in ensuring good use of public funds, and higher responsiveness to the needs of 'innovation consumers' through improved collaboration between public and private participants, including across national borders. A well-functioning agricultural innovation system is key to improving the economic, environmental and social performance of the food and agriculture sector. The long-term positive impact of agricultural R&D on productivity growth is well established, and technologies and practices can help improve the sustainability of natural resource use.

The statistical data for Israel are supplied by and under the responsibility of the relevant Israeli authorities. The use of such data by the OECD is without prejudice to the status of the Golan Heights, East Jerusalem and Israeli settlements in the West Bank under the terms of international law.

General innovation profile

Agriculture innovation is increasingly dependent on general innovation, through developments in Information and Communication Technology (ICT), biotechnology and nanotechnology, but also marketing innovation. A thriving innovation profile will ensure that general knowledge and specific knowledge in other fields, which are needed to develop and implement agriculture innovation are available, and that economic actors and society in general share an innovation culture.[1]

The **global Science, Technology and Innovation (STI) strategy** developed for 2009-20 commits to strengthen public research, improve science-industry collaboration and international linkages, strengthen human capital and improve governance (OECD, 2012). As recommended in a 2011 review of publicly funded research (Department of Industry, 2011), the Australian Research Committee (ARCom) was established to provide integrated and strategic advice on future research investments. It developed the *National Research Investment Plan*, which was released in 2012 and comprises:

- The objective of guiding Australian Government research investment;
- A *Framework* that facilitates the development of Australia's research capacity and capability to be responsive to the needs of all sectors including business;
- A set of *research investment principles*; and
- A statement of Strategic Research Priorities, which is currently under revision.

Regarding **STI policy governance**, the Department of Industry is responsible for industry, energy, resources, science and skills to drive economic growth, productivity and competitiveness.[2] Several councils were established to improve co-operation and provide policy advice. The *Framework of Principles for Innovation Initiatives* provides guidance to enhance consistency across the system and to improve the accessibility and efficiency of innovation initiatives across Australia. The major national science agencies are included in the Department of Industry portfolio, as are the Australian Research Council (ARC) and Intellectual Property Australia (OECD, 2012).

Figure 7.1. Science and innovation profile of Australia

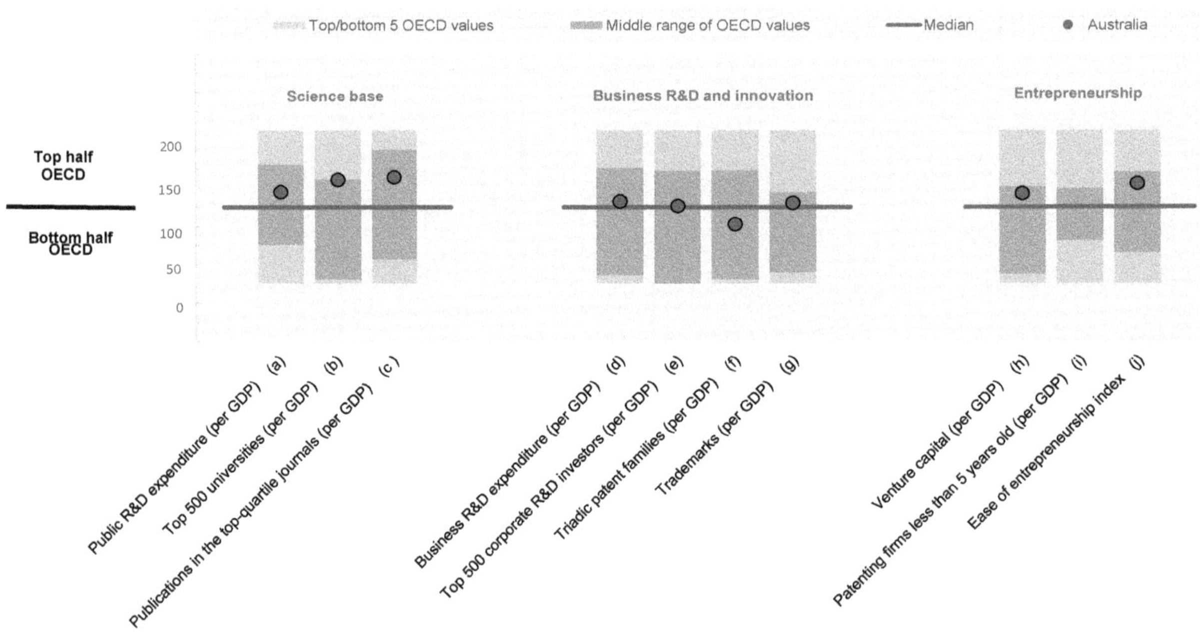

Source: OECD (2012), "Australia", in *OECD Science, Technology and Industry Outlook 2012*, OECD Publishing, Paris. DOI:. http://dx.doi.org/10.1787/sti_outlook-2012-35-en.

In assessing the **general innovation profile** of Australia, OECD (2012) finds that the public supply of knowledge is above the OECD average, as shown by its high public-sector expenditure on R&D in proportion to GDP, and the high international ranking of its universities and publication rates in top scientific journals (Figure 7.1). The public education system is also well-funded and Australia has strong skills base, in particular in science (Section 4.3). The workforce displays a high share of human resources in science and technology. Science and engineering degrees, as well as the number of researchers, are slightly above their OECD averages. Business expenditure on R&D is close to the OECD median, despite public incentives to private R&D, favourable conditions to entrepreneurship, strong ICT and scientific infrastructure, and good skills base. While trademarks per GDP are at the OECD median, patenting is below. This would indicate returns on science and investment are low, but it could also reflect the nature of research, and the composition of industries, with primary and resource-based industries accounting for a much larger share of business expenditure on R&D than the OECD median.

While the agricultural system has specific features and actors, it is integrated in the general innovation system through priority setting and contributions from general research organisations and universities to rural R&D and education.

Overview of agricultural innovation systems' actors and respective roles

Agricultural innovation systems involve a wide range of actors who enable, guide, fund, perform, implement, inform and facilitate innovation. The key players include policy-makers, researchers, teachers, advisors, farmers, private companies and consumers. Figure 7.2 shows the diversity of stakeholders in the Australian "rural"[3] innovation system, including funders, providers and end-users, while Figure 7.3 explains linkages between funders and providers through programmes.

R&D provision

The main providers of rural R&D in Australia are the Commonwealth Scientific and Industrial Research Organisation (CSIRO), state and territory governments, universities and private providers:

- CSIRO is the largest supplier of rural R&D in Australia.
- State and territory primary industry departments operate a geographically dispersed network of research institutes, experiment stations and extension services.
- Australian universities have had a strong role in agricultural research in Australia. Twelve universities, located in both metropolitan and rural areas, offer a degree course in agriculture or related agricultural areas.
- Private provision of rural R&D includes industry-owned providers; and large farming operations and multinational chemical and fertiliser companies.

Historically, CSIRO and the universities have been the main providers of basic research. However, in the past two decades, through increased partnerships described below, universities now conduct more project-focused, applied research.

Figure 7.2. Rural research and development system stakeholders

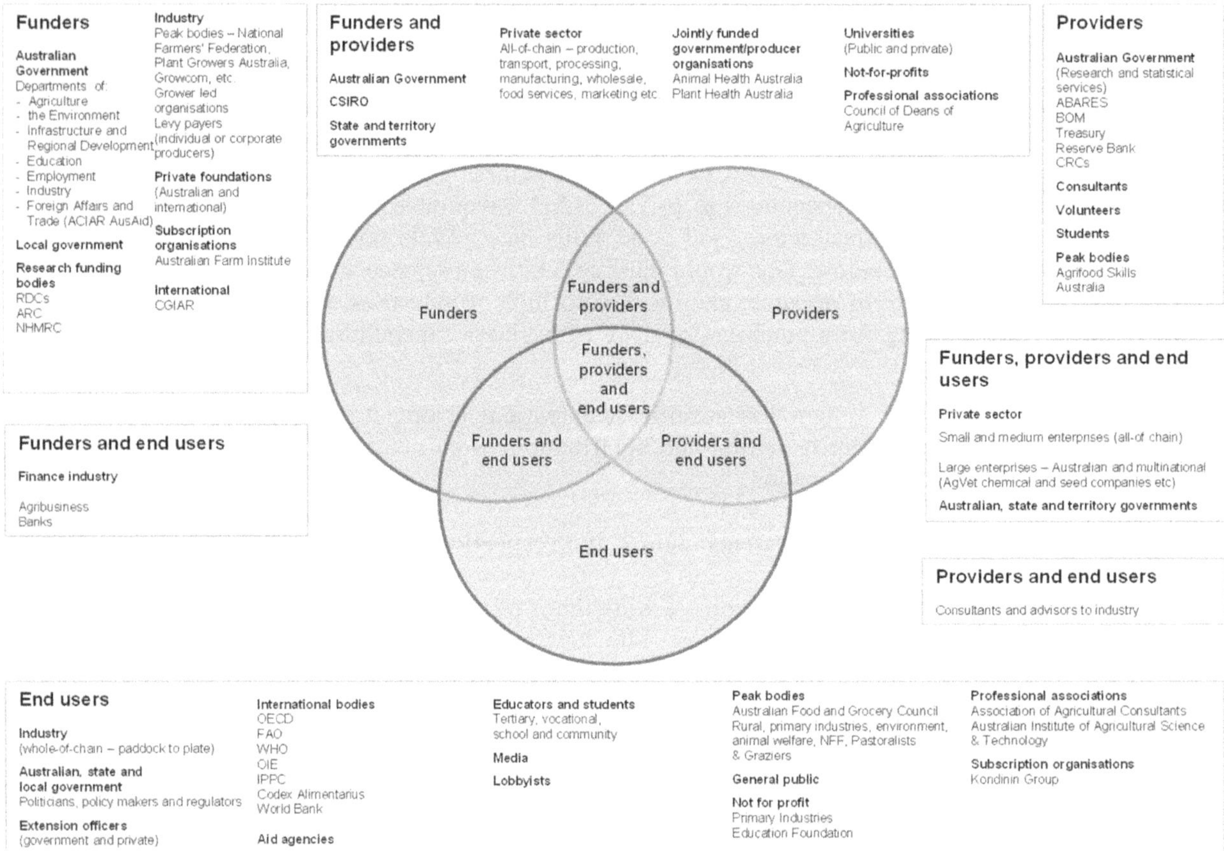

ACIAR: Australian Centre for International Agricultural Research; RDC: Research and Development Corporations; ARC: Agricultural Research Council; NHMRC: National Health and Medical Research Council; ABARES: Australian Bureau of Agricultural and Resource Economics and Sciences; BOM: Bureau of Meteorology; CRC: Cooperative Research Centres; RDC: Research and Development Corporation; OECD: Organisation for Economic Cooperation and Development; FAO: Food and Agriculture Organization of the United Nations; WHO: World Health Organization; OIE: World Organisation for Animal Health; IPPC: International Plant Protection Convention.

Source: Update from Australian Government (2012), Rural Research and Development Policy Statement, Department of Agriculture, Fisheries and Forestry, Canberra. http://www.daff.gov.au/__data/assets/pdf_file/0004/2176222/research-and-development-policy-statement.pdf.

Figure 7.3. Rural R&D funders, providers and programmes

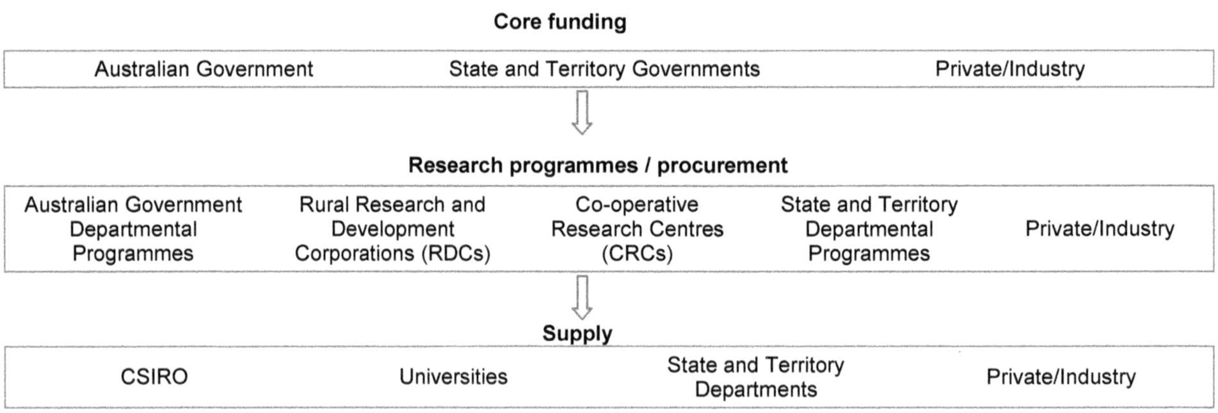

CSIRO: Commonwealth Scientific and Industrial Research Organisation.

Source: Productivity Commission (2011), *Rural Research and Development Corporations*, http://www.pc.gov.au/projects/inquiry/rural-research.

Extension

A wide diversity of public and private actors provide technical assistance and extension services to farmers. With state and territory governments reducing direct provision of extension services, there has been an increase in the number of private agronomists providing these services, as well as grower groups, not-for-profits and some joint public and private investment. In some industries, RDCs have taken on the extension role formerly provided by State and Territory Governments. Machinery, seed, fertiliser and agricultural veterinary chemical suppliers or companies also provide extension services to farmers pertaining to their products.

Governance

Governance ensures that government priorities are coordinated and communicated clearly, that progress is monitored and that policy outcomes and impacts are evaluated against objectives. The integration of the agricultural innovation system in the governance of general innovation ensures better use of public funds, and increased efficiency of innovation systems through the pooling of different expertise.

Setting of R&D priorities

The **Rural R&D Priorities for public investment** were reviewed and updated by the Australian Government in 2007, in consultation with state and territory governments, industry, and research funders and providers (DAFF, 2007). They were based on National Research Priorities established in 2002 and took into account the recommendations of several policy reviews, including the 2006 Corish Report (Agriculture and Food Policy Reference Group, 2006), which examined future directions in Australian Government policies and programmes impacting on the agriculture and food sector. The current government is reviewing its rural R&D priorities.

The Australian state and Northern Territory governments, rural R&D corporations, CSIRO, and universities are jointly developing the **National Primary Industries RD&E Framework**[4] to encourage greater collaboration and promote continuous improvement in the investment of Research, Development and Extension (RD&E) resources nationally. The Framework supports greater collaboration and coordination between the rural RD&E participants (including the Australian and state governments, rural R&D corporations, the CSIRO and universities); strengthens national research capabilities to better address sector and cross-sector issues; and focuses RD&E resources so they are used more effectively, efficiently and collaboratively, thereby reducing capability gaps, fragmentation and unnecessary duplication in primary industries RD&E (Box 11.1 in Productivity Commission, 2011,).

All 22 national RD&E strategies have been endorsed, and are in the implementation phase:

- Fourteen sectoral strategies (beef, cotton, dairy, fishing and aquaculture, forestry, grains, horticulture, new and emerging industries, pork, poultry, sheep meat, sugar, wine, and wool).
- Eight cross–sectoral strategies (animal welfare, biofuels and bioenergy, climate change water use in Australian agriculture, animal biosecurity, food and nutrition, plant biosecurity and soils).

Rural industries set their own investment priorities, guided by the National Research Priorities, the Rural R&D Priorities and the strategies developed under the National Primary Industries RD&E Framework. All bodies that receive public funding for rural R&D are expected to address the National Research Priorities and the Rural R&D Priorities in developing their investment strategies.

Evaluation

Evaluation is taking place at different levels: system, portfolio, industry, programme and project. Various tools and indicators are used to assess economic, environmental and social returns from investment in rural RD&E.

The Research and Development Corporations (RDC) model described below and the broader rural and national R&D policy environments have been reviewed periodically to identify and implement improvements to arrangements, and to ensure that the best value is achieved from public and private investments in rural RD&E. Key reports include:

- The recent Productivity Commission (2011) inquiry into the rural RDCs, which examined the effectiveness of the RDC model and the appropriateness of the current funding levels and arrangements.
- The Rural Research and Development Council's National Strategic Rural Research and Development Investment Plan (RRDC, 2011).
- The Productivity Commission (2007) research report on public support for science and innovation in Australia.
- An Industry Commission (1995) report that examined the effect of R&D on innovation, industry competitiveness and economic growth; and the performance of policies and programmes which influence R&D and innovation in Australia.
- Other commissioned reviews (Agriculture and Food Policy Reference Group, 2006; Cutler, 2008) [see Productivity Commission (2009) for a discussion of key findings from the Cutler Review (2008)].

The statutory funding agreements for each RDC require the RDC to be independently reviewed before instatement of a new agreement. In 2014 two individual RDCs were reviewed; an independent review of Horticulture Australia Limited, and a senate inquiry into Meat and Livestock Australia.

Stakeholders have identified a need for more information on the performance of Australia's rural RD&E system.[5] The Productivity Commission (2011) in its inquiry on the RDCs, and the Rural Research and Development Council in preparing its National Strategic Rural Research and Development Investment Plan (RRDC, 2011) highlighted that under the arrangements of that time, inadequate data impede governments at all levels in making strategic RD&E investment decisions. The Council also identified a need for regular examination of the performance of the rural R&D system, to inform strategic planning, guide improvements and demonstrate achievements. In addition, two wide-ranging, independent reviews of selected RDC projects were undertaken (Acil-Tasman, 2008, 2010).

A performance measurement and reporting framework was developed by the Rural Research and Development Council (2011) and refined by ABARES (Cuevas-Cubria et al., 2012). The framework uses a programme logic approach to link indicators of research inputs, outputs and outcomes (economic, environmental and social). ABARES assembled a set of indicators for regular reporting according to two criteria: suitability as measures of rural RD&E performance and ease of collection. For example, Category A indicators are those that ABARES considers should be collected and reported as part of a regular reporting framework. Table 7.1 lists category A indicators and provides examples of their potential uses.

While useful data are available in many areas, overall, ABARES found that some data limitations are likely to constrain full assessment of the rural RD&E system's performance. The coverage of input and output indicators appears sufficient to enable analysts to draw broad conclusions about the system's performance based on multiple indicators. In contrast, adequate data are not available to assess performance against implementation or non-economic social outcomes (Cuevas-Cubria et al., 2012).

Table 7.1. Category A indicators used to evaluate rural R&D

Indicator	Potential use	RD&E stage
Expenditure on rural R&D	To quantify financial inputs to the rural RD&E system	Input
Trends in completed tertiary award courses in rural-related fields	To assess the future capacity of the rural RD&E system	Input
Educational attainment of those employed in the rural sector	To analyse the capacity of rural business owners to adopt innovations	Input
Number and quality of Australian rural-related research outputs	To measure the knowledge created through rural research	Output
Trends in Australian registrations of rural-related patents and plant breeder's rights	To quantify new technologies and plant varieties created	Output
Number of rural firms engaged in innovative activity	To measure innovation adoption	Implementation
Trends in productivity growth for rural industries	To measure economic outcomes/efficiency improvements from RD&E	Outcome - economic
Return on RD&E investments	To measure the return from government investments in rural RD&E	Outcome - economic
Measures of environmental quality	To assess environmental outcomes from rural RD&E	Outcome - environmental

Source: Cuevas-Cubria, C. et al. (2012), *Measuring and reporting trends relating to the performance of Australia's rural RD&E system.*

Investing in innovation

The public sector continues to be the main source of funding for rural R&D, whether performed in public or private organisations. A wide variety of funding mechanisms can be used from direct spending on research projects and institutions to various forms of tax incentives. Business investment in R&D is normally driven by market demand, but governments also provide different kinds of incentives. Some, like R&D tax incentives, apply to the economy in general, while others are agriculture specific. In many countries, producer organisations and other non-governmental organisations also provide R&D funding. Knowledge infrastructure is a public good that can enable innovation; it includes ICT infrastructure and general purpose technologies as well as specific knowledge infrastructure such as databases and institutions.

R&D funding mechanisms

The Australian Government has a range of programmes, spread across several departments, which provide funding for rural R&D and shape the integrated and inclusive nature of the Australian agricultural innovation system.

In broad terms, the current arrangements used to provide public support for science and innovation involve: competitive funding, institutional funding, tax concessions and subsidies (Productivity Commission, 2007, p.28). Competitive funding arrangements involve:

- peer reviewed competitive grants — allocation is on the basis of perceived quality in response to researcher initiated proposals (ARC research grants);
- competitive tenders against predefined objectives — precise project objectives and outcomes are specified by the funding agency (such as RDC funded research); and

- other competitive grants and loans — all other competitive grants and loans for which project objectives are not specified in advance.

A significant proportion of Australian Government spending on rural R&D is conducted through **rural Research and Development Corporations (RDCs)**, within the Agriculture, Fisheries and Forestry portfolio. This is a co-investment model, established in 1989, whereby the RDCs procure rural R&D using funds collected from primary producers via statutory or voluntary levies, as well as funding provided by the government, generally on a matching basis up to a cap (Box 7.1).

Box 7.1. The rural Research and Development Corporation (RDC) model

The RDC model is the Australian Government's primary vehicle for funding rural innovation. Under the RDC model, the Australian Government matches industry R&D levy payments, dollar for dollar, with maximum matching contribution per year of 0.5% of an industry's gross value of production.

The co-investment model:

- Helps to ensure that all producers who benefit from research contribute to its cost, and to address free-rider problems that could lead to underinvestment in R&D of direct benefit to producers.
- Helps to elicit additional, socially valuable R&D, including where the benefits are either spread thinly across a wide range of industries, or mainly accrue to the wider community.
- Helps to ensure that public money is not spent on research of little practical value.
- Facilitates greater and faster uptake of research outputs.

In addition, it generates greater spending capacity. For instance, while the Grain RDC (GRDC) has an annual budget of around AUD 120 million, grains research organisations such as the Home Grown Cereals Authority (HGCA) in the United Kingdom and the Western Grains Research Foundation in Canada have budgets of around AUD 10 million and AUD 5 million respectively.

Source: Productivity Commission (2011), Rural Research and Development Corporations, http://www.pc.gov.au/projects/inquiry/rural-research.

In 2012, there were 15 RDCs that cover virtually all agricultural industries, as well as fisheries and forest and wood products (Figure 7.4). Many RDCs take a whole-of-value-chain approach to research. Rather than focusing only on on-farm issues, the RDCs also invest in areas such as quality and food safety management across value chains, and processing innovations.

Figure 7.4. The rural research and development corporation model

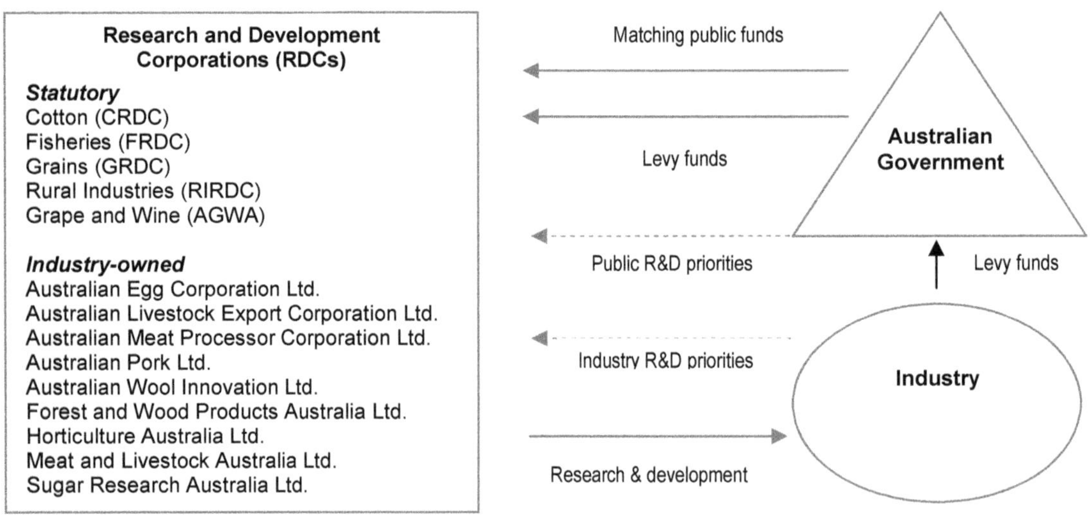

Source: ABARES.

The **Rural R&D for Profit Program**, administered by the Department of Agriculture, is providing an additional AUD 100 million in funding over four years to the rural RDCs, beginning in 2014/15. The Program aims to deliver cutting edge technology and applied research, with an emphasis on making the results accessible to primary producers.

The **Cooperative Research Centres (CRCs)**, administered by the Department of Industry, are partnerships between different research funders, suppliers and end users formed to undertake R&D in specific areas, with a particular emphasis on applied R&D. CRCs must include a university and an end user, with other possible partners including an RDC, the CSIRO, industry representative, or government organisation. CRCs receive public funding, through the Innovation, Industry, Science and Research portfolio, which must be matched by participants' cash and in-kind contributions, for a period of up to ten years via a competitive merit-based selection process. For 2014/15, there are 36 CRCs operating, seven of which are agriculture ones.[6]

The Department of Agriculture is responsible for funding innovative research and on–farm trials that reduce greenhouse gas emissions from the land sector and store carbon in the landscape, enhancing sustainable agricultural practices.

Much of the funding by state and territory governments is directed at in-house research conducted in state and territory research institutes and experimental stations, and related extension activities. In addition, state and territory governments contribute funding and or in-kind contributions to the CRCs and RDCs.

Composition of rural R&D funding

The public sector is the main source of rural R&D funding in Australia. Most of this comes through the Australian Government's levy matching funding of the rural RDCs, as core funding for the Commonwealth Science and Industrial Research Organisation (CSIRO) and competitive grants though the CRCs. Funding to universities and other Australian Government programmes is also provided (Productivity Commission, 2011).

In 2008/09 total funding for rural R&D was in the order of AUD 1.5 billion. The government share of this total funding was approximately 75% (Table 7.2), compared to an average share below 40% for the whole economy (OECD, 2013a, Table B1). The Australian Government contributed around two-thirds of total government funding. Australian Government programmes which provide funding for rural R&D are spread across several departments (Productivity Commission, 2011).

Around 25% of rural R&D in 2008/09 was funded by the private sector, although the share of private funding varies considerably across industry sectors (Productivity Commission, 2011). Downstream sectors in many rural industries also invest in R&D. The red meat processing, broiler and fruit canning sectors have all indicated their industry members make substantial in-house investments in R&D (Australian Meat Processor Corporation; Australian Chicken Meat Federation; and Canning Fruit Industry Council of Australia in submissions to Productivity Commission, 2011).

Industry levy payments to the RDCs, industry-owned research institutions, and state-based research organisations accounted for two-thirds of private funding for rural R&D, the remaining third coming from large commercial farming companies; and chemical, fertiliser and other agricultural supply companies (Table 7.2).

Table 7.2. Level and composition of agriculture R&D funding, 2008/09[1]

Organisation type	Funding Million AUD	Share %
Australian Government[2]		
Cooperative Research Centres	63	
Core funding for the CSIRO	193	
Core funding for universities[3]	118	
Research and Development Corporations (RDCs)	218	
Other departmental programmes[4]	114	
Foregone tax receipts arising from R&D tax concession	9	
Total Australian Government	*715*	*48*
State and Territory Governments		
Project-related budget allocations[5]	348	
Capital investment in R&D facilities	47	
Payments to other funders and suppliers	21	
Total State and Territory Governments	*416*	*28*
Private/industry		
Levy payments provided to RDCs	248	
Other (for which a tax concession is claimed)[6]	116	
Total Private/industry	*364*	*24*
Total	1 495	100

1. These data have been updated since the draft report. They do not include funding from royalties and other intellectual property income (on the basis that these have been generated by past funding from governments and private parties). Also, the data do not include in-kind contributions from the private sector, such as through the provision of land and facilities for experiments.

2. Only the portion of the budget assigned to rural R&D is included.

3. Estimated by applying the rural share of total university funding received from contestable sources and the portion of university students studying in agriculture-related areas to the three largest university block grants.

4. Includes programmes aimed at wider issues (such as climate change), programmes with no sector-specific focus and any one-off payments.

5. Includes rural R&D and associated extension funding for programmes facilitated within the primary industry department (or its equivalent). Any funding for rural R&D from State and Territory Government environment departments and the like is not included.

6. Calculated using tax concession data (including an estimate for concessions claimed for R&D on agricultural chemicals). Also includes payments made to the Australian Animal Health Laboratory.

Source: Productivity Commission (2011), *Rural Research and Development Corporations*, Report No.52,, http://www.pc.gov.au/projects/inquiry/rural-research.

Under the Primary industries Excise (Levies) Act 1999, the Primary industries Customs (Charges Act) 1999, and the Primary Industries Levies and Charges Collection Act (1991), levies are collected from processors by the Department of Agriculture and forwarded to the relevant RDC. In some industries processors are able to recover levies paid from relevant producers. In other industries processors are themselves liable:

- Red meat processors pay levies to the Australian Meat Processor Corporation on cattle and livestock that are processed.
- Timber mill operators pay the forest industries products levy to Forest & Wood Products Australia.
- Sugar mills pay 50% of the sugar cane levy to Sugar Research Australia Limited.

- Winemakers pay the wine grape levy if they own the product at the time it is made into wine, and the grape research levy if they own the product at the time of delivery for crushing (to the Australian Grape and Wine Authority).

Finally, royalties and other intellectual property income generate funding for further research. For example, by 2010, End-Point Royalties for wheat provided sufficient revenue to support all downstream commercial wheat breeding activities (Alston and Gray, 2013).

Trends in public R&D expenditure

From the late 1970s, growth in public agricultural R&D expenditure slowed (Figure 7.5). The average annual growth rate declined from around 7% a year between 1952/53 and 1977/78, to around 0.6% a year from 1977/78 to 2006/07. Moreover, research intensity (defined as the ratio of public R&D expenditure to agricultural gross domestic product) peaked at over 5% in 1977/78, before declining to 3% in 2006/07 (Sheng et al., 2011).

OECD data on government budget appropriations or outlays for R&D (GBAORD), which exclude other sources of funding for public R&D, show that public research intensity in agriculture more than double the average of all sectors. Around 2010, it was lower than public research intensity in Canada but close to levels found in the Netherlands or the United States (Figure 7.6).

Public-private R&D cooperation

The CRC programme is the main mechanism to support medium and long term end-user driven research collaboration. It aims to overcome the challenges with collaborative ventures involving research partners from diverse professional cultures (reviewed in Productivity Commission, 2007).

Figure 7.5. Real public expenditure on agricultural R&D and extension in Australia: 1952-53 to 2006-07

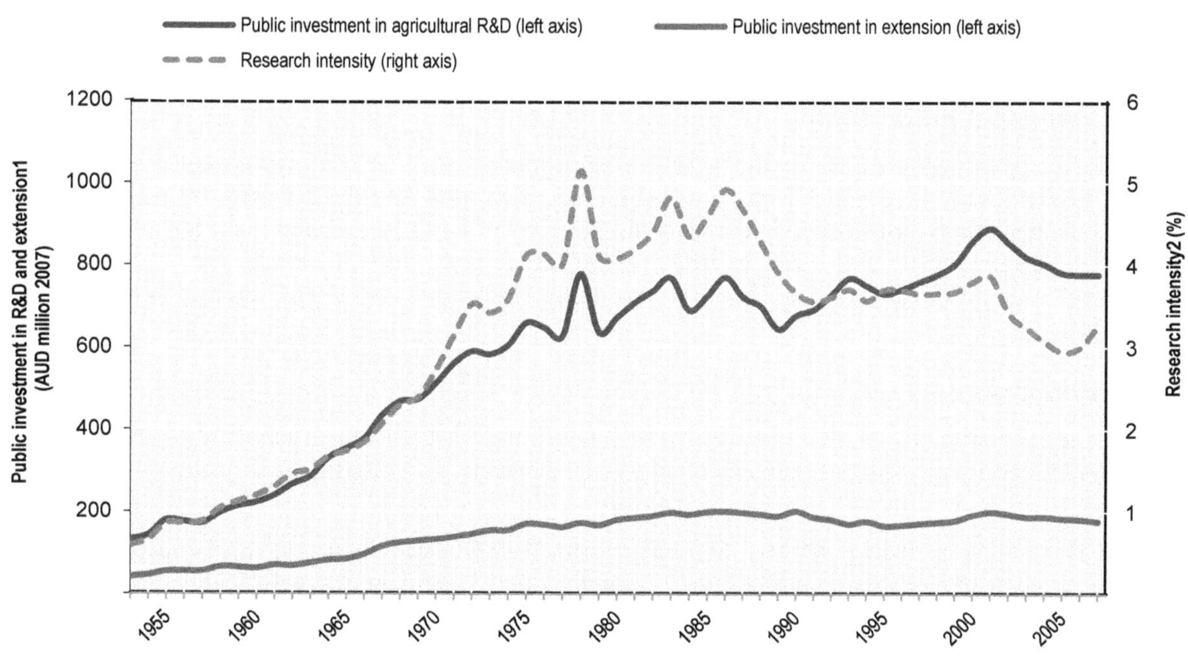

1. Public expenditure on agricultural R&D and extension includes expenditure by Australian, state and territory governments, and research institutions and universities. Funds from research and development corporations (excluding grower levies) and other external funders for agriculture (excluding research in fisheries and forestry) are also included.

2. Agricultural research expenditure as a percentage of agricultural GDP.

Source: Sheng, Y. et al. (2011), *Public investment in agricultural R&D and extension: An analysis of the static and dynamic effects on Australian broadacre productivity*, http://data.daff.gov.au/data/warehouse/pe_abares20110914.01/RR11.07_PubInvAgRandD.pdf.

Figure 7.6. Government budget appropriations or outlays for R&D on agriculture R&D as a share of agriculture value-added, 1995, 2010

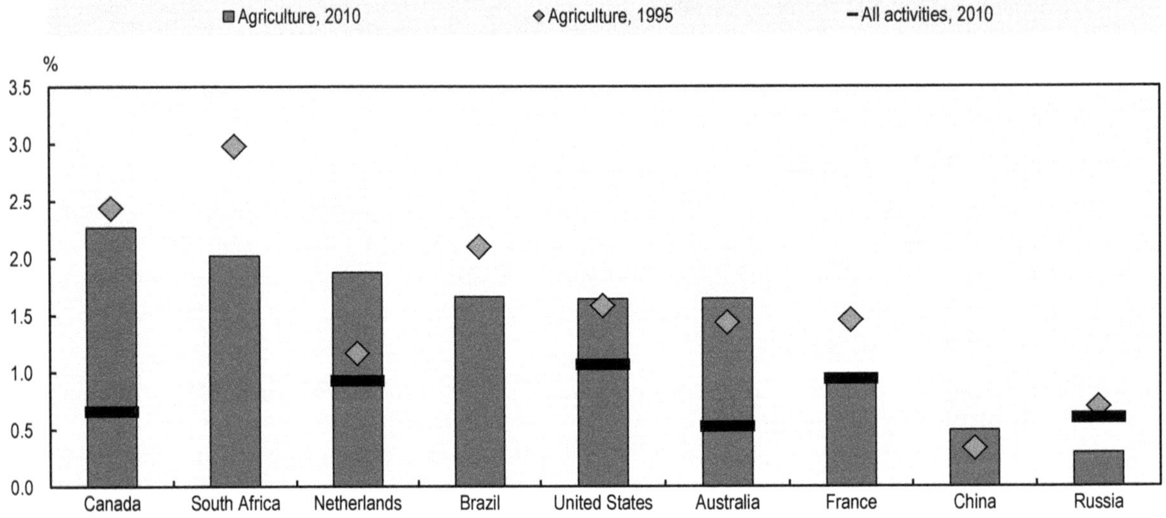

In 1988 classification changed from ISIC rev3 to rev4.

Source: OECD Research and Development Statistics and National Accounts, 2014; ASTI for non-OECD countries (Agricultural Science and Technology Indicators) IFPRI, 2014.

StatLink http://dx.doi.org/10.1787/888933250095

The CRC programme is, however, only suited to longer-term arrangements. There are complementary options for business collaboration with public sector research agencies and universities that could provide more nimble, less management-intensive, arrangements (Productivity Commission, 2007).

Public incentives to private R&D

Private researchers compete on the same ground as public ones for competitive funding described above. As explained in Section 3.4, tax concessions are also used to encourage private research. Tax concessions do not have a competitive element and include the R&D Tax Incentive for any eligible firm undertaking eligible R&D activities.[7] In 2008/09, the foregone tax receipts arising from the R&D tax concession accounted for a small share of Australian Government funding of agricultural R&D (Table 7.2).

Knowledge infrastructure

Public infrastructure for R&D includes research facilities at CSIRO and universities and research and statistics services in federal, state and territory Departments, as well as national collections held by CSIRO.

CSIRO has more than 6 500 staff located across 56 sites throughout Australia and overseas. It manages three National Research Facilities,[8] including the Australian Animal Health Laboratory at Geelong, which conducts research on emerging infectious disease threats and covers the livestock and aquaculture industries. CSIRO also has more than 30 other research facilities. Some of CSIRO's centres are of direct relevance for rural R&D. The High Resolution Plant Phenomics Centre (HRPPC) at CSIRO Plant Industry, for example, is developing with the Australian National University, the next generation research tools to probe plant function and performance both under controlled conditions and in the field. Recent advances in robotics, imaging and computing are used in applying these technologies and scaling them for analysing single plants to an entire ecosystem. Other CSIRO centres provide general knowledge, which helps advance rural research such as biodiversity or ICT.

CSIRO Collections[9] of interest to rural R&D include:

- Australian National Insect Collection (ANIC).
- Australian National Wildlife Collection.
- The Australian National Herbarium.
- Australian Tree Seed Centre.

In addition, CSIRO maintains a National Soil Archive, which is Australia's primary resource for conserving soil specimens, and make them accessible for future research.

Twelve universities, located in the metropolitan and rural areas, have agriculture or related agricultural activities.

State and territory primary industry departments operate a geographically dispersed network of experiment stations and extension services close to local producers. Partly because of the large capital cost of refurbishing outdated infrastructure, this network has been contracting. Stakeholders have raised concerns that the contraction of resources for R&D infrastructure would jeopardise future R&D performance (Productivity Commission, 2011, Box 11.2). The National Primary Industries RD&E Framework initiative is likely to lead to both further rationalisation of the network and much greater specialisation in research across the jurisdictions.

ABARES[10] is a research bureau within the Department of Agriculture. It provides economic and scientific research, commodity forecasts, statistics and other analysis to inform government and private sector decision-makers on significant issues affecting Australia's agriculture, fisheries and forestry industries.

Knowledge management is an important aspect of government's role in AIS, as it has impacts on all areas. Primary Industries Steering Committee (PISC) agencies agreed to store their publicly available research outputs in their own digital repositories and expose these via the web using standard formats which enable the repositories to be accessed and retrieved by specialised external web search services. The Steering Committee is now called Agricultural Senior Officials Committee (ASOC).

The Australian Bureau of Statistics (ABS) and the Australian Bureau of Agricultural and Resource Economics and Sciences (ABARES) are undertaking a review of the National Agricultural Statistical System. Final recommendations are expected to be publicly released on the ABS website before the end of 2014.

Fostering knowledge flows: The role of Intellectual Property Rights

Intellectual property rights (IPRs) are of growing importance in fostering innovation, which increasingly requires collaboration and exchanges of knowledge.

In Australia, there are standard and innovation patents, covered under the Patents Act 1990 (Productivity Commission, 2007, Appendix N). The main points of difference between these two types of patents are the level of inventiveness needed to gain certification and the maximum level of protection. The innovation patent, replacing the petty patent from 2001, requires a lower level of inventiveness and is aimed at protecting innovations which are not greatly different from existing technologies, but still have a significant commercial value. In line with the lower inventive threshold, the innovation patent is cheaper to obtain, requires less certification testing and has a faster approval process. However, it is also shorter and gives a lower level of protection compared to the standard patent (Table 7.3). Standard patent fees in Australia are lower than the OECD average and among the lower tier of OECD countries.

Table 7.3. Standard versus innovation patents, as at 1 February 2007

	Standard patent	Innovation patent
Maximum duration (years)	20	8
Application[1] (AUD)	770	440
Maintenance cost[2] (AUD)	7 830	900
Inventiveness or innovativeness test	A new invention not obvious to an expert in that field	A substantial contribution to the working of an invention
Main industry users	Life science and ICT industries	Consumer goods, mining and transport industries

1. All patent fees are to change from 1 March 2007. Assumes online application, no Australian intellectual property examination report (for standard patents) and no third party requests (for innovation patents).

2. Assumes fees paid for maximum life of patent. A standard patent can be extended past 20 years for AUD 1 200 per year.

Source: Productivity Commission (2007), "Public Support for Science and Innovation", http://www.pc.gov.au/__data/assets/pdf_file/0016/37123/science.pdf.

Over the years, Australia has further strengthened protection of IPRs by expanding its international commitments (WTO, 2011). According to the World Economic Forum, Intellectual Property protection in Australia is relatively high as it is in many countries of similar level of development (Figure 7.7). Patent protection increased significantly in the 1990s, but remains lower than in the United States. Australia became a member of the International Convention for the Protection of New Varieties of Plants (UPOV) in 1989 and signed the latest 1991 Act in 2000. As a result plant variety protection increased to a level close to that in the United States.

The signature of UPOV91 and the creation under the Plant Breeder's Act 1994 of a system of end-point royalties (EPRs), which enable owners of Plant Variety Rights to collect revenue in Australia, attracted private interest and involvement in seed and plant breeding (Alston and Gray, 2013).

There are a number of domestic public and private institutions that implement and support the IP system:

- **IP Australia** is an independent agency of the Department of Industry and is the central regulatory and administering body for IP in Australia. It administers Australian IP, processes IP claims, collects IP application and maintenance fees, maintains the Australian IP register, provides information, liaises with international bodies and undertakes policy research and analysis.
- **Technology transfer offices:** The commercialisation and technology transfer offices of universities and public sector research agencies offer advisory services and expertise to the research community within these institutions. According to Uniquest,[11] while the specific service charter may vary from office to office, they generally bridge 'the gap between basic research performance in universities and the development of new products and processes in industry'. Other aspects of their operation include liaising with business to sell patents, forming start-up companies and encouraging business engagement of research services. These agencies are becoming increasingly common as commercialisation strategies are employed to ensure research output utilisation. An example of University technology transfer offices is Uniquest, which oversees the IP portfolio of the University of Queensland and, recently also, the University of Wollongong.

Figure 7.7. Intellectual Property Protection

A. Patent Protection Index, 1960-2010
Score 1-5 (best)

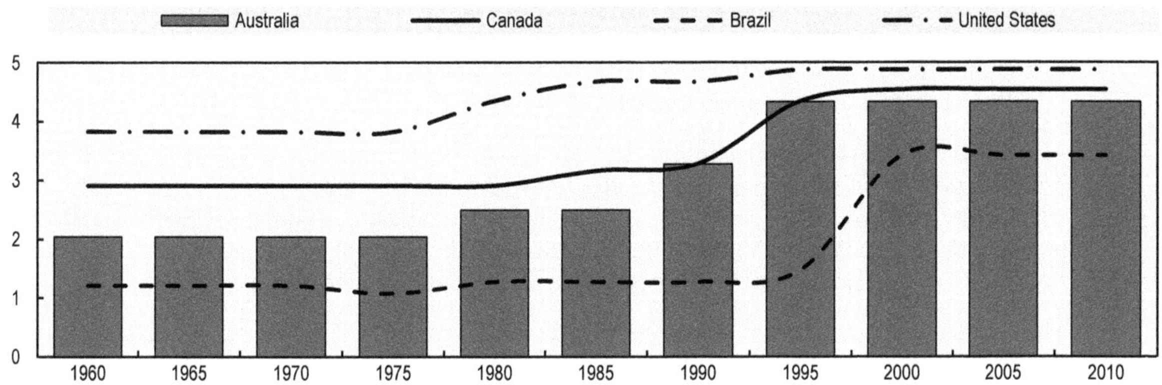

Sum of indices for duration, enforcement, loss of rights, membership and coverage.
Source: Unpublished update to Park, W. G. (2008), "International Patent Protection: 1960-2005.

B. WEF Intellectual Property Protection Index
Score 1-7 (best)

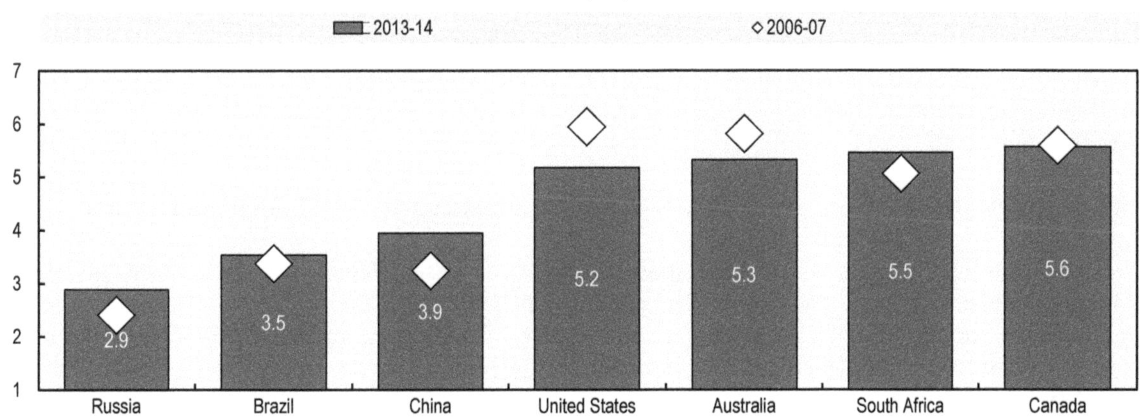

Source: World Economic Forum (2013), *The Global Competitiveness Report 2013-2014: Full data Edition*, Geneva 2013. http://reports.weforum.org/the-global-competitiveness-report-2013-2014/#=.

C. Plant Variety Protection Index
Score 1-5 (best)

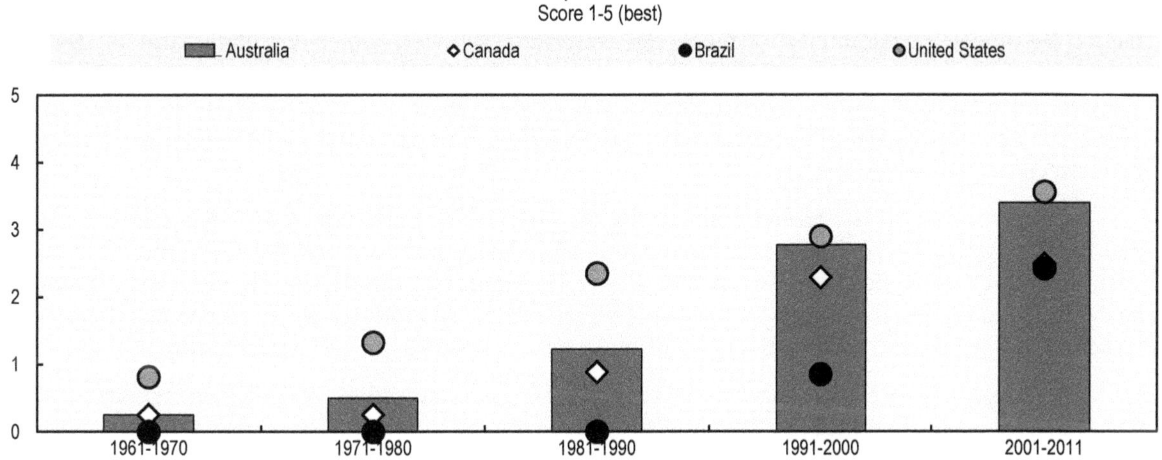

Campi et al. (2013), *Intellectual property protection in plant varieties: A new worldwide index* (1961-2011), http://hdl.handle.net/10419/89567.

StatLink http://dx.doi.org/10.1787/888933250107

The Advisory Council on Intellectual Property and the Intellectual Property Research Institute of Australia (IPRIA) both undertake research on a wide range of topics in the field of IP.

- The Advisory Council is an independent body appointed by the Australian Government to advise the Minister for Industry on intellectual property matters and the strategic administration of IP Australia. It has performed a number of reviews on certain aspects of the IP system, resulting in policy changes.
- The IPRIA[12] is a national centre for multi-disciplinary research on the law, economics and management of IP. It is based at the University of Melbourne and is run jointly by the Faculty of Law, the Faculty of Economics and Commerce, and the Melbourne Business School. It focuses on making contributions to the general understanding of the optimal settings for IP policy and how these interact with levels of innovation.

- The Intellectual Property and Competition Review Committee (IPCRC) was established by the Australian Government to review the impact of IP policy on competition.
- The Australian Institute of Commercialisation monitors IP-related issues in relation to commercialisation.

Principles for IP Management have been developed within the context of the National Primary Industries RD&E Framework.[13]

Many creators utilise advisory services, such as technology transfer offices or patent attorneys to assist in the process of applying to a patent. While Universities and CSIRO apply mainly for international patents, a majority of CRCs' patents are issued for the Australian market (Productivity Commission, 2007, Table N.2).

Facilitating knowledge flows and linkages within the national agricultural innovation system

Reinforcing linkages across participants in the AIS (researchers, educators, extension services, farmers, industry, NGOs, consumers and others) can help match the supply of research to demand, facilitate technology transfer, and increase the impact of public and private investments. Partnerships can also facilitate multi-disciplinary approaches that can generate innovative solutions to some problems.

Many factors affect the **diffusion of innovation** among farms, including structural characteristics such as age of farmers, education level, or management and financing capacity. This section discusses two aspects: 1) the relevance of innovation results, i.e. their ability to respond to farmers' demand for solutions that for example generate additional or more stable income, save labour or preserve scarce natural resources; 2) agricultural training and extension services, which facilitate farmers' access to improved technology and knowledge enabling them to adapt to changing circumstances and solutions. They can also facilitate farmer participation in innovation networks and the ability to formulate specific demands. The importance of these services is revealed by farmers' willingness to pay for them, either collectively or individually.

Within the Australian agricultural innovation system, the RDCs and CRCs facilitate cooperation and technology transfer. Through co-financing and industry involvement, the RDC model is built to ensure research respond to industry needs (Box 7.1). Moreover, the RDCs provide some commodity-based targeted extension services to their levy payers. The CRCs are partnerships between the research community, funders and end-users and focus on specific, applied areas of interest to the industry.

The supply of extension services is very diverse and interactive, and evolves rapidly to match demand. There are differences by commodity sector, with the grains industry relying almost entirely on private consultants or farmer groups, while livestock grazing industries rely more on services provided through RDCs (OECD, 2014).

Direct provision of extension services by the public sector delivered by the state and territory governments to primary producers has decreased over time and remaining services tend to focus on biosecurity and public good areas such as natural resource management and climate change mitigation.

Production-focussed extension services are provided through the rural RDCs at a programme and/or project level, thus with public co-funding (DAFF, 2013c). In addition, farmer-led grower alliance extension type services emerge. They are often funded by RDCs, such as the Grains Research and Development Corporation and Dairy Australia. Farmer-led organisations are also involved in regional delivery of Australian Government funded sustainable resource management practices and climate change mitigation practices (DAFF, 2013a, 2013b).

Extension can be funded as part of RD&E activities or government programmes. Research projects funded under the *Rural R&D for Profit* initiative will be required to include a pathway to adoption for their project outcomes. The initiative will also have an element specifically devoted to strengthening Australia's rural research extension systems. The Australian government's public good *Caring for Our Country and Carbon Farming Initiatives* includes extension programmes to facilitate uptake of natural resource management and climate change mitigation government initiatives by farmers (DAFF, 2013a; DAFF, 2013b). The *Extension and Outreach Program*, administered by the Department of Agriculture provides technical information and support to assist farmers, land managers and their key influencers to participate in land sector emissions management activities and the Carbon Farming Initiative.

While services on a user-pay basis are generally available, concerns have been raised that the reduction in commitment of extension by state and territory government primary industries agencies may have created a service gap especially in remote areas and/or specialised small-scale industries, which are not covered by a RDC. Although agri-business has expanded their services, advice given by chemical or fertiliser companies often has a commercial imperative (Zhou, 2013). Another concern is the change in service delivery model from government extension officer individual on-farm visits to extension delivered through adult-learner/group service delivery (Zhou, 2013). Although farmers often have a preference for face-to-face on-farm extension delivery, the diversity of production systems, reduction in free government extension services and distances travelled between farms often preclude this type of service now (AIA, 2013).

Strengthening international co-operation on agricultural innovation

International co-operation on agricultural research and development offers universal benefits. While this is generally true given the public good nature of many innovations in agriculture, it is particularly the case where global challenges are being confronted (as in the case of responding to climate change) and when initial investments are exceptionally high. The benefits of international co-operation for national systems stem from the specialisation it allows and from international spill-overs. In countries with limited research capacity, scarce resources could then focus on better taking into account local specificities.

Australia has a long history of involvement in international rural RD&E collaboration. R&D linkages to other countries are built around Australia's strong participation in international conferences, Australian Centre for International Agricultural Research (ACIAR) programmes, Australia's involvement in the international agricultural research centres of the Consultative Group on International Agricultural Research (CGIAR), as well as partnerships formed by the CSIRO and universities.

In 2008–09, ACIAR provided RD&E funding to 15 international agricultural research centres and to projects across Papua New Guinea and the Pacific, South-east Asia, South Asia, North Asia and Southern Africa. Of the RDCs, the Grains Research and Development Corporation formed a substantive relationship with two of the CGIAR Centres (ICARDA and CIMMYT).[14]

A recent example of Australian international activity is the International Knowledge Based Bio-Economy (KBBE) Forum. This quadrilateral grouping of the European Commission (EC), Australia, Canada and New Zealand seeks to enhance current bilateral project-based cooperation with strategic and research and innovation policy-related discussions on bio-economy challenges and opportunities; and scientific cooperation at multi partner level. Australia is leading the healthy food work stream, which is one of four work streams including biotechnologies for biorefineries and bio-based materials (Canada), fish and aquaculture (EC) and sustainable agriculture (New Zealand).

Another example is the Global Research Alliance on Agricultural Greenhouse Gases[15] established in 2009 to bring countries together to find ways to grow more food (and build more climate-resilient food production systems) without growing greenhouse gas emissions; to find ways to reduce the emissions intensity of agricultural production and increase its potential for soil carbon sequestration, while enhancing food security; to improve understanding, measurement and estimation of agricultural emissions; and improve farmers' access to agricultural mitigation technologies and best practices. Australia and France are leading the Soil Carbon and Nitrogen Cycling cross cutting research group.

Measuring the performance of the agricultural innovation system

Overall progress to create and adopt relevant innovations can usefully be monitored, using various indicators of R&D outcomes, diffusion and impact.

R&D outcomes

The number of patents in a country is a proxy for R&D outcomes, but it is not a comprehensive indicator of the performance of the innovation system, as not all innovation are patented, not all patents are being used, and other IPR systems exist for plant varieties. In addition, numbers should be complemented with indicators of patent quality, which are being developed at OECD (2013b).

According to agricultural patent applications filed under the Patent Co-operation Treaty (PCT), one of the systems which protects inventions in all signatory countries, Australia's contribution to world agriculture patents covering specific agricultural inputs, agricultural sciences and food processing innovations is modest, and much lower than its contribution to world scientific publications (Table 7.4). This difference may reflect the applied nature of the research on agriculture conducted in Australia, the dominance of public research, and its focus on the national market.[16] Both patent and bibliographic contributions are lower than in OECD countries like Canada, Japan, France, Germany and the United Kingdom (Figures 7.A1.1 and 7.A1.3).

As in most countries, the majority of agriculture-related patents are in food processing, but compared to many developed countries, the share of patents in agricultural sciences is relatively high (Figure 7.A1.1). Agricultural sciences are more dominant in terms of bibliographic references on agriculture. Australia's share of agricultural science publications and citation in the total world output has decreased in the last 15 years (Figure 7.8). This reflects the strong increase in contributions by emerging economies countries, as the number of agriculture publications by Australian authors increased by 6% per year between 1996 and 2012 (compared to around 28% in the BRIICS[17] and around 5% in the OECD area).

R&D cooperation

Close to a fourth of agriculture-related patents and half of scientific publication originate from cooperation between Australian and foreign research (Table 7.5). The rate for publication is slightly below the OECD average and to that found for Canada, for example. Although there is no systematic rule, higher rates are found in smaller or less-developed countries and lower ones in large economies like the United States and Russia (Figure 7.A1.5). In terms of patents, the rate of cooperation is in the lower range (Figure 7.A1.2).

Table 7.4. R&D outcomes, 2006-11

	Australia	Brazil	Canada	United States	OECD average	OECD total
Agri-food specialisation: Agri-food as a share of country's total (%)						
Patents	7.4	11.0	6.0	6.8	5.6	
Publications	10.6	19.4	8.7	6.7	9.4	
Citations	10.8	15.5	8.3	6.3	11.9	
Country's contribution to world agri-food output (%)						
Patents	0.5	0.2	0.6	10.8	0.7	27.9
Publications	3.3	4.7	3.7	18.3	2.0	68.9
Citations	2.9	1.2	4.1	27.2	..	48.4

Source: OECD Patent Database, January 2014; SCImago. (2007). SJR — SCImago Journal & Country Rank. http://www.scimagojr.com.

Figure 7.8. Evolution of scientific output and impact in agricultural sciences, 1996-2012

Percentage of world output

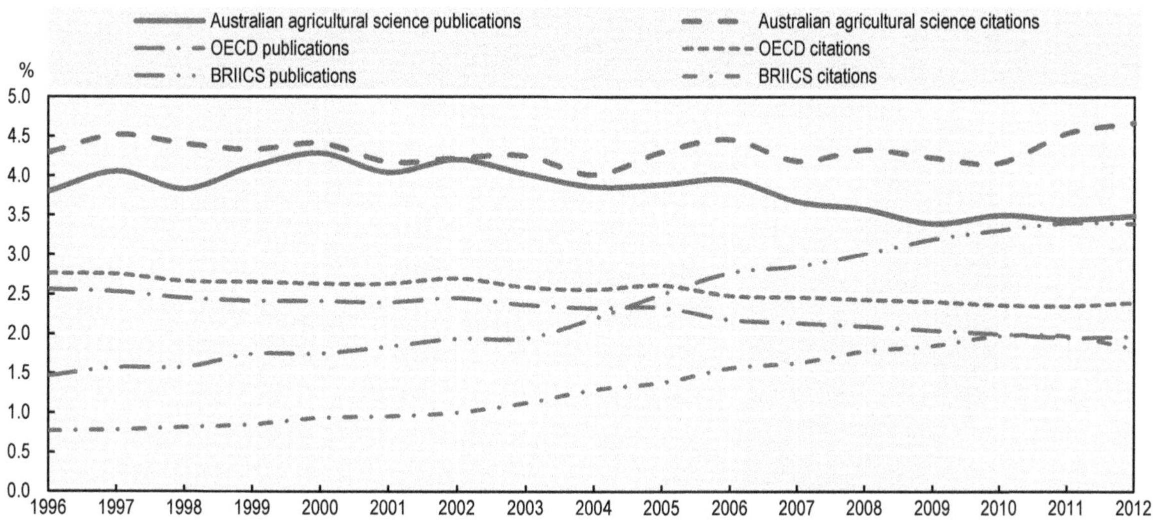

Agricultural science include Scopus journal classifications: agronomy and crop science, animal science and zoology, aquatic science, ecology/evolution/behavior systematics, forestry, horticulture, insect science, plant science and soil science, and miscellaneous agriculture/biological sciences.

Source: SCImago. (2007). SJR — SCImago Journal & Country Rank. Retrieved 19 March 2014, from http://www.scimagojr.com.

StatLink http://dx.doi.org/10.1787/888933250119

Table 7.5. R&D co-operation, 2006-11

Agri-food outputs with co-authors as a share of total agri-food outputs (%)

	Australia	Brazil	Canada	United States	OECD average
Patents	23.1	29.7	29.7	14.3	11.8
Publications	47.3	22.3	48.9	36.4	50.8

Source: OECD Patent Database, January 2014; SCImago. (2007), *SJR — SCImago Journal & Country Rank*, http://www.scimagojr.com.

Contribution to productivity growth

A recent ABARES report (Gray, Oss-Emer and Sheng, 2014) considers how changes to Australia's agricultural and economy-wide policy settings have contributed to agricultural productivity growth. Among other things, it concludes that "investments by the Australian, state and territory governments in rural R&D and extension have had a significant effect on agricultural productivity. ABARES research found that past investments in broadacre R&D and extension by Australian governments (with the latter mainly provided by state and territory government primary industries departments) have generated internal rates of return that could be as high as 28% and 47% a year, respectively (Sheng et al., 2011).

In addition to innovations generated domestically, Australian agriculture has also benefited from knowledge and technology developed overseas. ABARES research found that spill-overs from foreign R&D (proxied by investment from the United States) have accounted for average broadacre TFP growth of around 0.63 percentage points annually (Sheng et al., 2011). Moreover, the relative contributions of foreign and domestic research (including domestic extension) to broadacre TFP growth have been roughly equal, suggesting that Australian agriculture relies heavily on international research spill-overs" (Figure 7.9).

Figure 7.9. Relative contributions of public R&D and extension to annual broadacre TFP growth (%)

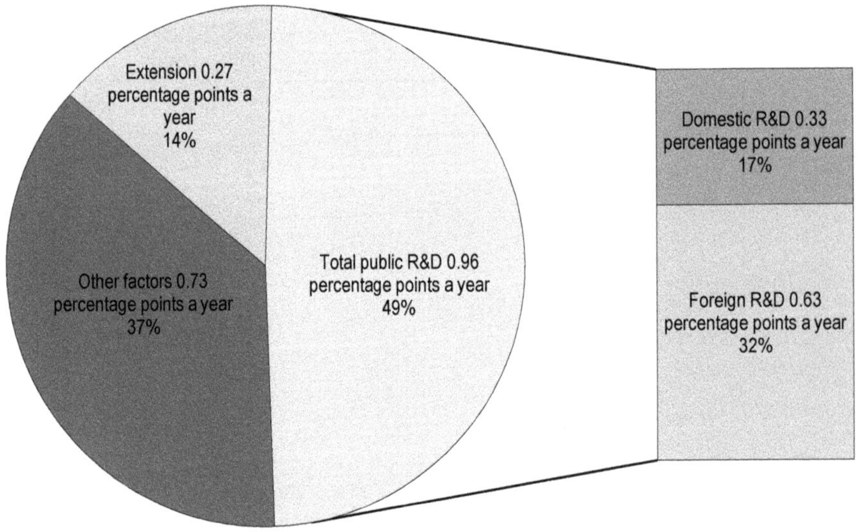

Source: Sheng, Y. et al. (2011), Public investment in agricultural R&D and extension: An analysis of the static and dynamic effects on Australian broadacre productivity.

Returns on R&D investment

In Australia, cost-benefit analyses are frequently used to measure the economic, environmental and social returns of R&D investment. Overall, they show large returns from investment in agricultural RD&E, even when conservative estimates are used for benefits that are difficult to quantify (OECD, 2014). Sheng et al. (2011) found that internal rates of returns to public investment over the period 1952/53 to 2006/07 were 28% for R&D and 47% for extension. They also found that R&D strategies that invest over the long-term to increase knowledge stock resulted in higher returns than strategies that invest over the short-term. In contrast, they found that short-term investments in public extension yields higher returns over the short-term than over the long-term. An inquiry into the RDC system (Productivity Commission, 2011) considered more specifically the distribution of public and private benefits arising from public investment in R&D through RDCs. The returns to the public were seen as modest relative to the share of government funding in the system.

Notes

1. This section draws on the response by ABARES to the OECD framework questionnaire and the response to a questionnaire on Agricultural innovation systems, available at: http://www.oecd.org/tad/agricultural-policies/48251416.pdf, which provided background information to the OECD AKS Conference held in Paris in 2011 (See OECD web site www.oecd.org/agriculture/policies/innovation). The proceedings of the conference also include a section on Australia's approach to rural research, development and extension (Grant, 2012). The information has been updated to reflect recent developments up to October 2014.

2. www.innovation.gov.au/innovationframework.

3. In Australian policy documents as in this report, "rural" R&D means "agriculture and food-related" R&D.

4. See framework website at: http://www.npirdef.org/framework.

5. The following is from Cuevas-Cubria *et al.* (2012) and Australian Government (2012). More detail on the above questions can be found in Cuevas-Cubria et al. (2012).

6. For further information about the CRC programme, see www.crc.gov.au.

7. For more information, see R&D Tax Incentive http://www.business.gov.au/grants-and-assistance/innovation-rd/RD-TaxIncentive/Pages/default.aspx and Chapter 10 in Productivity Commission (2007).

8. CSIRO research facilities: http://www.csiro.au/Organisation-Structure/National-Facilities.aspx.

9. CSIRO Collections: http://www.csiro.au/Organisation-Structure/National-Facilities.aspx.

10. See ABARES website: http://www.daff.gov.au/abares/Pages/Default.aspx.

11. UniQuest is the main commercialisation company of The University of Queensland.

12. IPRIA: www.ipria.org.

13. See IP Principles: http://www.npirdef.org/files/resourceLibrary/resource/26_PISCIntellectualPropertyManagementPrinciples.pdf.

14. Center for Agricultural Research in the Dry Area (ICARDA) http://www.icarda.org/; International Maize and Wheat Improvement Center (CIMMYT) http://www.cimmyt.org/en/.

15. Global Research Alliance on Agricultural Greenhouse Gases: http://www.globalresearchalliance.org/.

16. A patent can also be registered in various national systems such as IP Australia, the European Patent Office (EPO) or the US Patent and Trademark Office (USPTO).

17. The BRIICS comprise Brazil, Russian Federation, India, Indonesia, China and South Africa.

References

Acil-Tasman (2008), *Measuring Economic, Environmental and Social Returns from Rural Research and Development Corporations' Investment*, Acil-Tasman Pty. Ltd. Canberra. www.ruralrdc.com.au.

Acil-Tasman (2010), *Impact of Investment in Research and Development by the Rural Research and Development Corporations*, Acil-Tasman Pty. Ltd. Canberra. www.ruralrdc.com.au.

Agriculture and Food Policy Reference Group (2006), *Creating our Future: Agriculture and Food Policy for the Next Generation,* report to the Minister for Agriculture, Fisheries and Forestry, Canberra, February. http://www.agriculture.gov.au/ag-farm-food/food/publications/creating_our_future_agriculture_and_food_policy_for_the_next_generation .

Ag Institute Australia (2013), "Future of agricultural extension workshop", 8 April, Canberra (unpublished).

Alston, J.M. and R.S. Gray (2013), "Wheat Research Funding in Australia: The Rise of Public-Private-Producer Partnerships", *EuroChoices,* Vol. 12, No. 1, pp. 30-35.

Australian Government (2012), Rural Research and Development Policy Statement, Department of Agriculture, Fisheries and Forestry, Canberra. Available at: http://www.daff.gov.au/__data/assets/pdf_file/0004/2176222/research-and-development-policy-statement.pdf .

Campi, M. and A. Nuvolari (2013), "Intellectual property protection in plan varieties: A new worldwide index (1961-2011)", *LEM Working Paper Series,* No. 2013/09 http://hdl.handle.net/10419/89567.

Cuevas-Cubria, C, Gibbs, C, Nossal, K, Gray, EM, Oss-Emer, M, Lawson, K & Davidson, A (2012), *Measuring and reporting trends relating to the performance of Australia's rural RD&E system*, ABARES report to client prepared for the Agricultural Productivity Division, Department of Agriculture, Fisheries and Forestry, Canberra, June. http://daff.gov.au/SiteCollectionDocuments/abares/publications/clientreports/measuring-and-reporting-trends-relating-to-the-performance-of-australias-rural-rd-and-E-system.pdf.

Cutler, T (2008), *Venturous Australia: building strength in innovation*, Report on the Review of the National Innovation System, Australian Government, Canberra. http://apo.org.au/node/2842.

DAFF (2013a), *Extension and Outreach*, at http://www.daff.gov.au/climatechange/carbonfarmingfutures/extensionandoutreach, viewed 20 May.

DAFF (2013b), *Regional Delivery*, at http://www.nrm.gov.au/funding/regional/index.html, viewed 20 May.

DAFF (2013c), Research and Development Corporations, at http://www.daff.gov.au/agriculture-food/innovation/research_and_development_corporations_and_companies, viewed 20 May.

DAFF (2007), Rural Research and Development Priorities, Department of Agriculture, Fisheries and Forestry, Canberra. http://www.daff.gov.au/agriculture-food/innovation/priorities.

Department of Industry (2011), *Focusing Australia's Publicly Funded Research Review - Maximising the Innovation Dividend: Review Key Findings and Future Directions*, October 2011. http://www.industry.gov.au/research/Pages/FocusingAustraliasPubliclyFundedResearch.aspx.

Grant, A. (2012), "Australia's approach to rural research, development and extension" in OECD (2012), *Improving Agricultural Knowledge and Innovation Systems: OECD Conference Proceedings*, OECD Publishing, Paris. DOI:. http://dx.doi.org/10.1787/9789264167445-en.

Industry Commission (1998), *Microeconomic Reforms in Australia: a Compendium from the 1970s to 1997*, Research Paper, Australian Government Publishing Service, Canberra, January. http://www.pc.gov.au/research/completed/microeconomic-reform-compendium-1970-1997.

Industry Commission (1995), *Research and Development*, Report No. 44, AGPS, Canberra.

OECD (2014), "Australian Primary Industries Research, Development and Extension System: Its evolution, Features and Impact on Green Growth Objectives" [COM/TAD/CA/ENV/EPOC/RD(2014)7/FINAL].

OECD (2013a), *Agricultural Innovation Systems: A Framework for Analysing the Role of the Government*, OECD Publishing, Paris. DOI: http://dx.doi.org/10.1787/9789264200593-en.

OECD (2013b), *OECD Compendium of Agri-Environmental Indicators*, OECD Publishing, Paris. DOI: http://dx.doi.org/10.1787/9789264181151-en.

OECD (2012), "Australia", in *OECD Science, Technology and Industry Outlook 2012*, OECD Publishing, Paris. DOI: http://dx.doi.org/10.1787/sti_outlook-2012-35-en.

Park, W.G. (2008), "International Patent Protection: 1960-2005", *Research Policy*, No. 37, pp 761-766, http://nw08.american.edu/~wgp/res_policy08.pdf.

Productivity Commission (2011), Rural Research and Development Corporations, Report No.52, Final Inquiry Report, Canberra. Available at: http://www.pc.gov.au/projects/inquiry/rural-research.

Productivity Commission (2009), Trade & Assistance Review 2007-08, Annual Report Series, Productivity Commission, Canberra, May. http://www.pc.gov.au/research/recurring/trade-assistance/tar0708.

Productivity Commission (2007), Public Support for Science and Innovation, Research Report, Productivity Commission, Canberra. Available at: http://www.pc.gov.au/projects/study/science/docs/finalreport.

RRDC (2011), National Strategic Rural Research and Development Investment Plan, Rural Research and Development Council, Department of Agriculture, Fisheries and Forestry, Canberra, March. http://www.agriculture.gov.au/ag-farm-food/innovation/national-strategic-rural-research-and-development-investment-plan.

Sheng, Y., E.M. Gray, J.D. Mullen and A. Davidson (2011), *Public investment in agricultural R&D and extension: an analysis of the static and dynamic effects on Australian broadacre productivity*, Australian Bureau of Agricultural and Resource Economics and Sciences, Canberra. http://data.daff.gov.au/data/warehouse/pe_abares20110914.01/RR11.07_PubInvAgRandD.pdf.

World Economic Forum (2013), *The Global Competitiveness Report 2013-2014*: Full data Edition, Geneva 2013. http://reports.weforum.org/the-global-competitiveness-report-2013-2014/#=.

WTO (2011), *Trade Policy Review: Australia—Report by the Secretariat (Revision)*, Trade Policy Review Body, WT/TPR/S/244/Rev.1, 18 May, World Trade Organization.

Zhou, Z-Y. (2013), *Developing successful agriculture: an Australian case study*, CAB, Cambridge.

Annex 7.A1.

Background Indicators of R&D Outcomes

Figure 7.A1.1. Agriculture patents applications filed under the Patent Co-operation Treaty (PCT), 2006-11

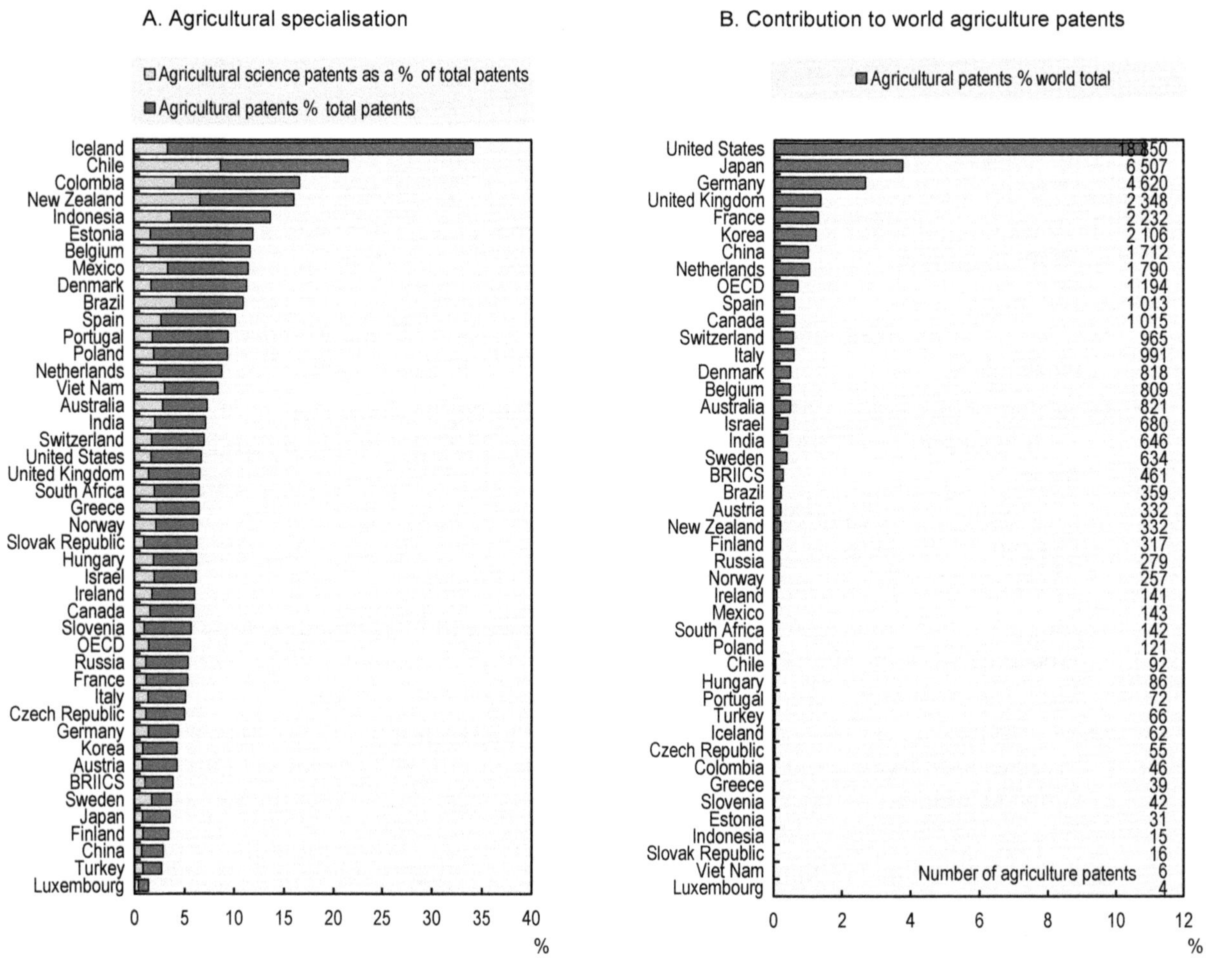

Agriculture includes patents from IPC classes A01, A21, A22, A23, A24, B21H 7/00, B21K 19/00, B62C, B65B 25/02, B66C 23/44, C08b, C11, C12, C13, C09K 101/00, E02B 11/00, E04H 5/08, E04H 7/22, G06Q 50/02.

Patent counts are based on the priority date (first filing of the patent worldwide), the inventors country of residence, using fractional counts.

Source: OECD Patent Database, January 2014.

StatLink http://dx.doi.org/10.1787/888933250128

Figure 7.A1.2. Agriculture patents with a foreign co-inventor filed under the Patent Co-operation Treaty (PCT), 2006-11

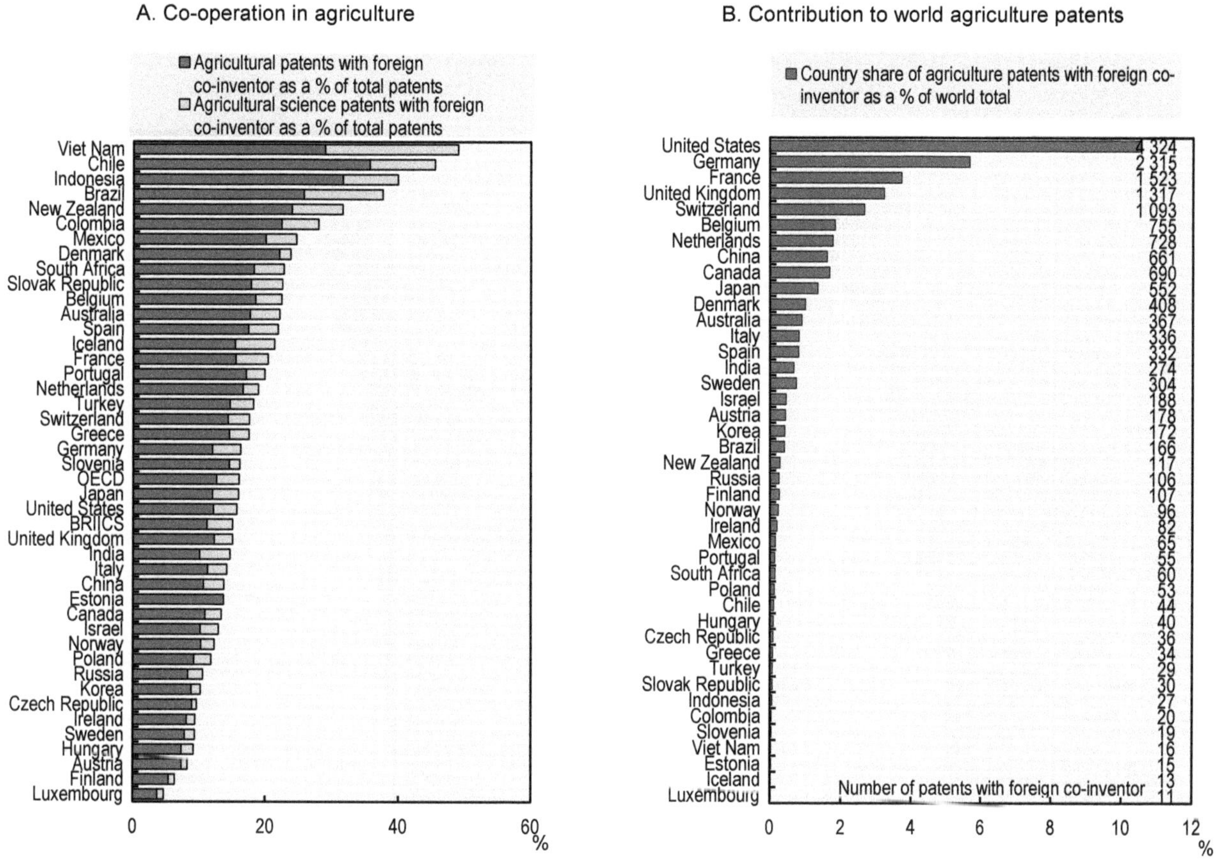

Agriculture includes patents from IPC classes A01, A21, A22, A23, A24, B21H 7/00, B21K 19/00, B62C, B65B 25/02, B66C 23/44, C08b, C11, C12, C13, C09K 101/00, E02B 11/00, E04H 5/08, E04H 7/22, G06Q 50/02.

Patent counts are based on the priority date (first filing of the patent worldwide), the inventors country of residence, using fractional counts. EU28 and BRIICS totals exclude intra-zone co-operations.

Source: OECD Patent Database, January 2014.

StatLink http://dx.doi.org/10.1787/888933250132

Figure 7.A1.3. Agriculture publications, 2007-12

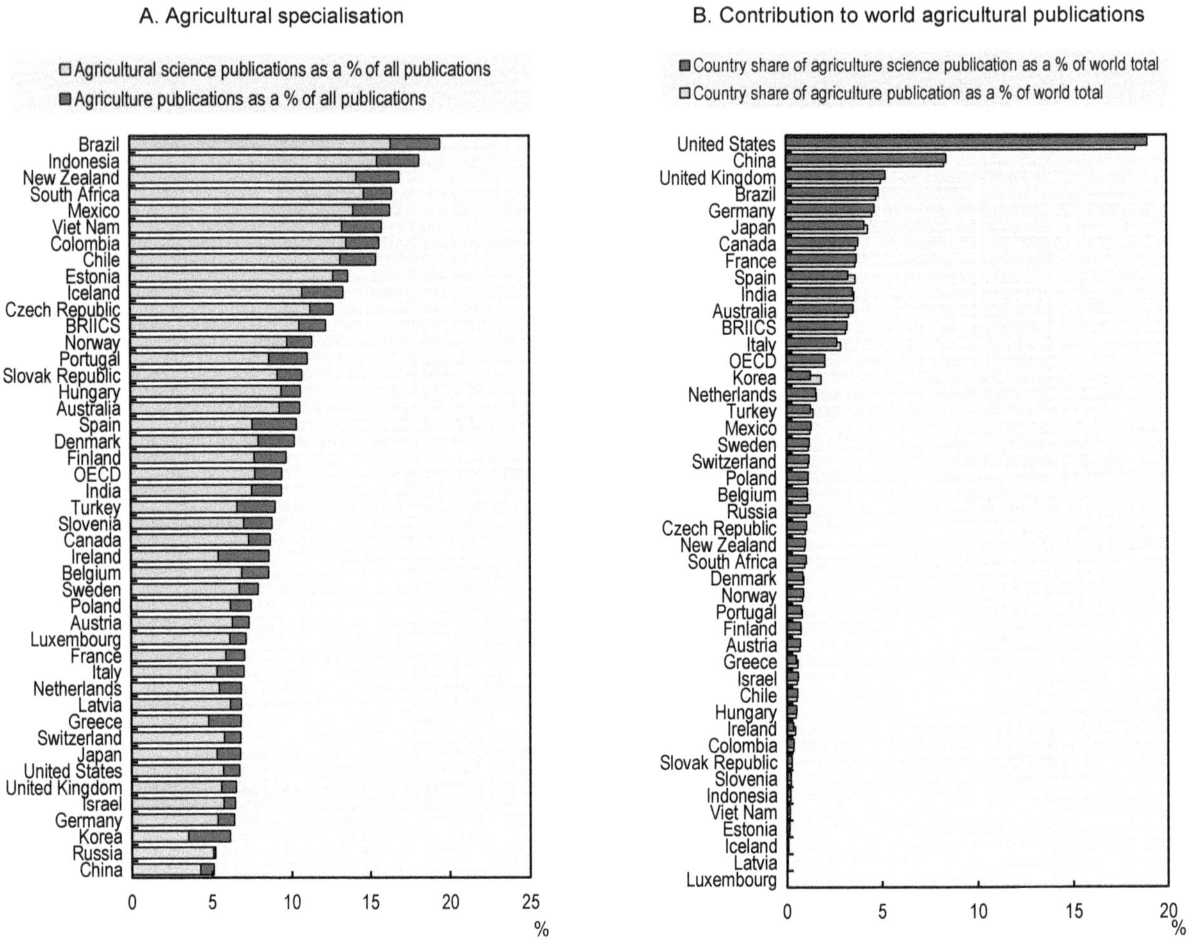

Agricultural sciences include Scopus journal classifications: agronomy and crop science, animal science and zoology, aquatic science, ecology/evolution/behaviour systematics, forestry, horticulture, insect science, plant science and soil science, and miscellaneous agriculture/biological sciences.

Agriculture publications include food sciences.

Source: SCImago. (2007). SJR — SCImago Journal & Country Rank. Retrieved March 2014 from http://www.scimagojr.com.

StatLink http://dx.doi.org/10.1787/888933250142

Figure 7.A1.4. Agriculture citations, 2007-12

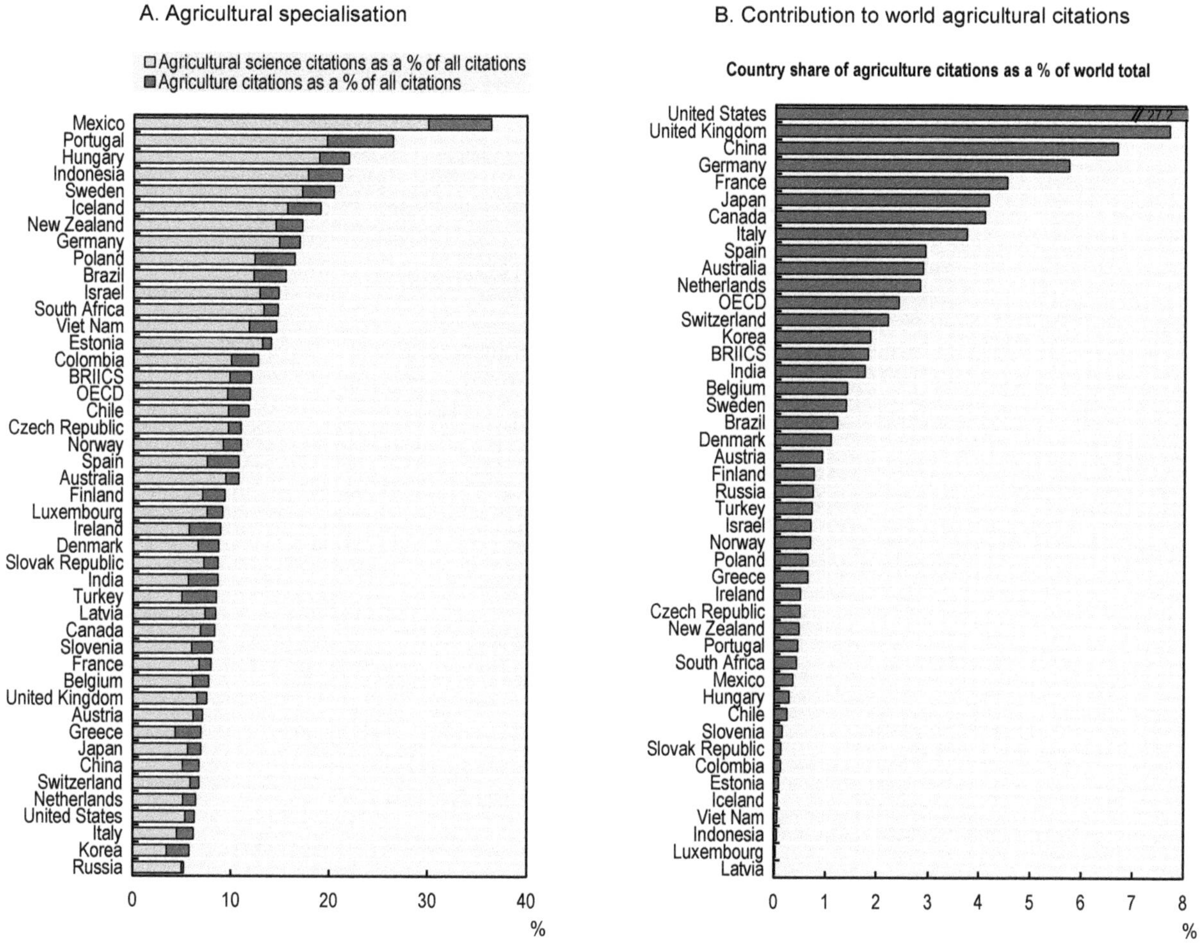

Agricultural sciences include Scopus journal classifications: agronomy and crop science, animal science and zoology, aquatic science, ecology/evolution/behaviour systematics, forestry, horticulture, insect science, plant science and soil science, and miscellaneous agriculture/biological sciences.

Agriculture citations include food sciences.

Source: SCImago. (2007). SJR — SCImago Journal & Country Rank. Retrieved March 2014, from http://www.scimagojr.com.

StatLink http://dx.doi.org/10.1787/888933250155

Figure 7.A1.5. International collaboration, 2007-12

Percentage of documents with collaborating authors in foreign country

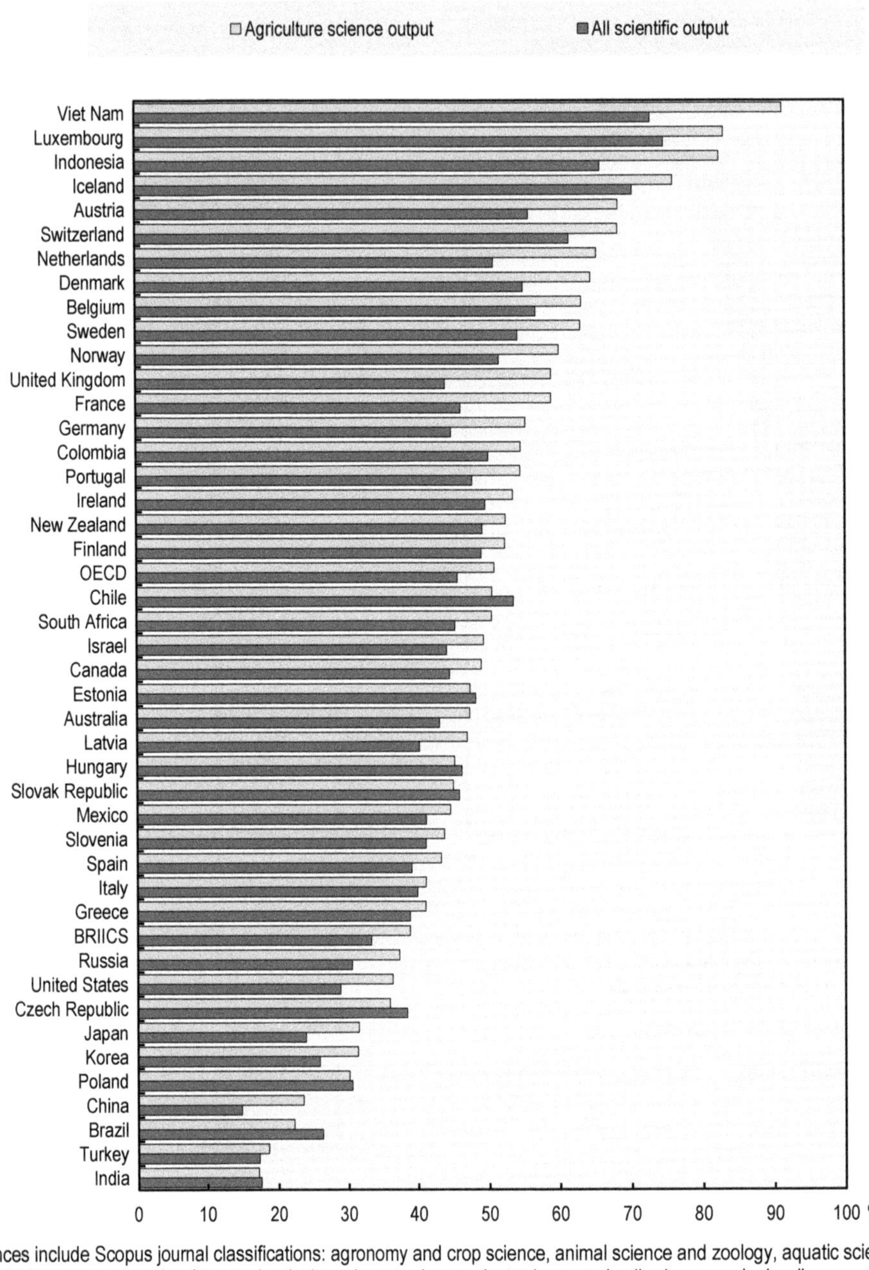

Agricultural sciences include Scopus journal classifications: agronomy and crop science, animal science and zoology, aquatic science, ecology/evolution/behaviour systematics, forestry, horticulture, insect science, plant science and soil science, and miscellaneous agriculture/biological sciences.

Source: SCImago. (2007). SJR — SCImago Journal & Country Rank. Retrieved March 2014, from http://www.scimagojr.com.

StatLink http://dx.doi.org/10.1787/888933250162

ORGANISATION FOR ECONOMIC CO-OPERATION AND DEVELOPMENT

The OECD is a unique forum where governments work together to address the economic, social and environmental challenges of globalisation. The OECD is also at the forefront of efforts to understand and to help governments respond to new developments and concerns, such as corporate governance, the information economy and the challenges of an ageing population. The Organisation provides a setting where governments can compare policy experiences, seek answers to common problems, identify good practice and work to co-ordinate domestic and international policies.

The OECD member countries are: Australia, Austria, Belgium, Canada, Chile, the Czech Republic, Denmark, Estonia, Finland, France, Germany, Greece, Hungary, Iceland, Ireland, Israel, Italy, Japan, Korea, Luxembourg, Mexico, the Netherlands, New Zealand, Norway, Poland, Portugal, the Slovak Republic, Slovenia, Spain, Sweden, Switzerland, Turkey, the United Kingdom and the United States. The European Union takes part in the work of the OECD.

OECD Publishing disseminates widely the results of the Organisation's statistics gathering and research on economic, social and environmental issues, as well as the conventions, guidelines and standards agreed by its members.

OECD PUBLISHING, 2, rue André-Pascal, 75775 PARIS CEDEX 16
(51 2015 13 1 P) ISBN 978-92-64-23832-9 – 2015

www.ingramcontent.com/pod-product-compliance
Ingram Content Group UK Ltd.
Pitfield, Milton Keynes, MK11 3LW, UK
UKHW051240300926
15471UKWH00015B/1396

9 789264 238329